Human, All Too (Post)Human

Human, All Too (Post)Human

The Humanities after Humanism

Edited By Jennifer Cotter, Kimberly DeFazio, Robert Faivre, Amrohini Sahay, Julie Torrant, Stephen Tumino, and Rob Wilkie

LEXINGTON BOOKS
Lanham • Boulder • New York • London

Published by Lexington Books
An imprint of The Rowman & Littlefield Publishing Group, Inc.
4501 Forbes Boulevard, Suite 200, Lanham, Maryland 20706
www.rowman.com

Unit A, Whitacre Mews, 26-34 Stannary Street, London SE11 4AB

British Library Cataloguing in Publication Information Available

Library of Congress Cataloging-in-Publication Data

Names: Cotter, Jennifer, editor.
Title: Human, all too (post)human : the humanities after humanism / edited by Jennifer Cotter, Kimberly DeFazio, Robert Faivre, Amrohini Sahay, Julie Torrant, Stephen Tumino, and Robert Wilkie.
Description: Lanham : Lexington Books, 2016. | Includes bibliographical references and index.
Identifiers: LCCN 2016011884 (print) | LCCN 2016013043 (ebook) | ISBN 9781498505734 (cloth : alk. paper) | ISBN 9781498505741 (Electronic)
Subjects: LCSH: Sociobiology. | Human behavior. | Humanities. | Biopolitics. | Civilization.
Classification: LCC HM628 .H86 2016 (print) | LCC HM628 (ebook) | DDC 304.5--dc23 LC record available at http://lccn.loc.gov/2016011884

∞™ The paper used in this publication meets the minimum requirements of American National Standard for Information Sciences Permanence of Paper for Printed Library Materials, ANSI/NISO Z39.48-1992.

Printed in the United States of America

Table of Contents

Acknowledgments

The following chapters are published in the present volume with permission from *The Red Critique*:

Chapter 1: "The New Class Common-Sense: Biopolitics, Posthumanism, and Love" by Jennifer Cotter was published in an earlier version titled "Bio-Politics, Transspecies Love, and/as Class Commons-Sense" in *The Red Critique* 14 (Fall/Winter 2012).

Chapter 2: "Loving Transnationalism: Spiritualizing Class in *House of Sand and Fog*" by Amrohini Sahay is a revised version of her essay titled "Transnational Healing in 'Karma Capitalism': Spirit and Class in *House of Sand and Fog*" published in *The Red Critique* 12 (Winter/Spring 2007).

Chapter 3: "'Theory Too Becomes a Material Force': Militant Materialism or Messianic Matterism?" by Stephen Tumino was published in an earlier version under the same title in *The Red Critique* 15 (Winter/Spring 2014).

Chapter 4: "Mind Over Matter and Other Posthumanist Feminist Tales" by Julie P. Torrant is a revised version of her essay titled "It Is Time To Give Up Liberal, Bourgeois Theories, Including New Materialist Feminism, and Take Up Historical Materialist Feminism for The 21st Century" published in *The Red Critique* 15 (Winter/Spring 2014).

Chapter 8: "The 'Event-al' Logic of Disaster: On 'Left' Extinctionism" draws on arguments previously published in an editorial titled "Disaster Theory" in *The Red Critique* 14 (Fall/Winter 2012).

Introduction

Posthumanism and the Evacuation of Critique

Jennifer Cotter, Kimberly DeFazio, Robert Faivre, Amrohini Sahay, Julie Torrant, Stephen Tumino, and Rob Wilkie

> *"Labour. . . is the prime basic condition for all human existence and this to such an extent that, in a sense, we have to say that labour created man himself."*
>
> Frederick Engels

Most theorists today treat posthumanism and the posthumanities—"decentering the human in favor of a turn toward the nonhuman, understood variously in terms of animals, affectivity, bodies, organic and geophysical systems, materiality, or technology"[1] —as the turn to encompass all turns. This is because in the dominant posthumanities the abandonment of the "human-centeredness" that sets boundaries between the human and the nonhuman is said to open the potential for creating a truly democratic planetary multiplicity without a center that exceeds all limits, whether ontological or epistemological. Against the "speciesism" of the "Anthropocene," posthumanism proposes life as *becoming-in-common*—what Jacques Derrida frames as "*l'animot*,"[2] Donna Haraway calls "transspecies encounter value,"[3] Bernard Stiegler names "technogenesis,"[4] Manuel DeLanda writes as "flat ontology,"[5] Bruno Latour describes as "practical metaphysics,"[6] Timothy Morton announces as "geophilosophy,"[7] and Rosi Braidotti, "a monistic ontology."[8] Posthumanism, in other words, functions as the spiritual sign of a community-to-come that moves beyond all social binaries and "unites not through con-vergence, con-version, or con-fusion, but rather through di-vergence, di-version, and dif-fusion."[9] The "community to come" is, in all its multiple mediations and philosophico-tropic idioms, an ecotheosophy of the dissolving of the fundamental binary of class which underpins capitalism.

Human, All Too (Post)Human: The Humanities after Humanism is an intervention into the posthumanism turn. We argue that posthumanism does not represent the death knell of humanism that announces the advent of the nonhuman common, as it is widely represented, but instead is humanism's latent side: the difference of humanism with itself as it adjusts to the heightening class contradictions in the shift from analog to digital capitalism. Humanism and posthumanism, in other words, are cultural representations of property relations that normalize wage labor, the main obstacle to social transformation. To go beyond humanism and the dominant posthumanities, we argue, requires a critique of the "spiritual aroma"[10] of both and the role they play in naturalizing the world as it is, as the way it will always be.

Following Nietzsche, the arche-theorist of posthumanism, the contemporary posthumanities treats itself as the voice of a post-speciesist ethics, cut free from a humanist modernity that linguistically divides the human from the animal and creates hatred for that which, in turn, becomes "animal."[11] Humanism's valorization of the human as the moral, conscious subject of free will that is the bearer of a universal essence, and its referential theory of language, according to which language, as interrelated signs, correspond to the empirical reality outside it, are seen by posthumanism as incapable of addressing the relation of human and nonhuman life and the emerging social realities. More specifically, the posthumanities views humanism's centering of the human and its theory of language as imposing arbitrary constructs and divisions and that, as a result, maintain the artificial boundaries between the human and the nonhuman (e.g., animals, nature, technology, objects) in contemporary life. Severing humans and language from their material relations, posthumanism ultimately creates a cultural imaginary in which human subjects are placed beyond conventional cultural and epistemological boundaries and divisions. Which is another way of saying that posthumanism spiritualizes the material relations produced by class.

Posthumanism, for instance, rejects humanism's worshiping of the industrious and self-determined ("free") individual, but disconnects these ideas from the relations of wage labor that are the material basis of humanism. By treating humanism as a way of thinking that needs to be negated by new thinking, posthumanism simply substitutes a new ethics of commonality among human and nonhuman life without addressing the material relations that, in the class interests of a few humans, systematically put all other human and nonhuman life at risk. In replacing the "selfish" ("free") human with a cooperative transspecies, it does not address the social relations within which the selfish and the cooperative develop and acquire their (class) value. Humanist modernity, in other words, is not merely an artificial regime constraining the vitality of "life." It is a relay of the social relations of industrial ("analog") capitalism, which required referential modes of thinking conducive to "mass production" and the Fordist mode of accumulation of surplus value and which fostered a more unified notion of subjectivity to suture class contradictions. The globally integrated social relations of digital (postindustrial) capitalism, which exploit labor through increasingly more complicated, high-tech forces of production, depends on updated notions of subjectivity, language, and understanding. That is, it requires subjects with a planetary consciousness of plurality and more complex theories of language for which there is never any single reliable referent but an irreducible multiplicity of referents. Through this framework, today's increasingly crisis-ridden empirical reality appears as in excess of all rational thought and language and thus as beyond class analytics and class struggle.

Relations between the human and the nonhuman, the natural and the technological, and the environmental and the social, are, to put what we have said differently, fundamentally shaped by the material relations of society, which under all stages of capitalism remain rooted in the appropriation of surplus value in the working day. But, whereas the dominant posthumanities treats historical conditions more in terms of "entanglement" and the "enmeshed" networks in which the boundaries between humans, animals, objects, and technology fade in and out of reality, we argue that such concepts actually erase the material structures of human and nonhuman life. They offer modes of cultural resistance to the growing alienation of daily life and appeal to the growing need to connect diverse events in global capitalism, but they disappear the exploitative relations between people that actually shape and explain humans' relation with technology and nonhuman life. It is not "technology" in itself that has harmed the environment and "decentered" existence, but technology as it is produced under for-profit social—i.e., *class*—relations of production that has deepened the environmental crisis and exacerbated the class struggle across all boundaries. Posthumanism naturalizes wage labor by depicting issues of climate change, environmental collapse, and economic crisis in apocalyptic terms: as aleatory, unrepresentable, and exceeding explanation. This fascination with apocalyptic thinking in the posthumanities is, more often than not, a means of displacing the social relations of production which shape what events become "disasters" and why. In contrast, what is necessary today, we argue, is the end of alienation and estrangement in the "negation of the negation" that is socialism, the "*actual* phase necessary for the next stage of historical development in the process of human emancipation and rehabilitation."[12]

This book thus offers a different reading of the posthumanism turn by implicating the posthumanities in the dialectics of capital and labor. The posthumanism turn's aggressive politics of affirmation, which equates affirmation with "life itself" and reduces critique to a "thanatopolitics" that fetishizes the negative and thus the negation of life,[13] has become a powerful paradigm for denying the negation of capitalism represented by the critique of its "root" in exploited and alienated labor. Posthumanism's moral rejection of the "arrogance" of humanism and its claim to an awareness of the devastating techniques of humanist "intentionality," not only for humanity but especially for the nonhuman environment in which both are inextricably "entangled," is an eco-ethics "without a subject" that reinscribes the spiritual agency traditionally ascribed to humans in Western theology to a transcendent idea of "materiality without materialism and even perhaps without matter."[14] It affirms the nonhuman as "wholly other"[15] to all that exists, materially and ideally, as the excess of the conceptual. What is thus proposed as a radical departure from the instrumental reason and humanist binaries of the Enlightenment that oppose the "world in itself" (*res extensa*) to the "world for us" (*res cogitans*) is the "world as occulted."[16] The consequence of such arguments is perhaps most clearly articulated by Eugene Thacker when he says about the occulted world that, "regardless of how much knowledge we produce about it, [it] always retains some remainder"[17] that cannot be explained by the social, and thus, cannot ground the emancipatory project of "finding a new or improved version of the world-for-us."[18] In the dominant posthumanities this spiritual common-ism (like Kant's *sensus communis*) represents the immanent secret of the world believed to exist above and beyond the social division of labor which militates against critique of the existing world so as to change it.

The dominant posthumanism, we argue, takes as its starting point the alienation of social life under capitalism which results from private property and, under the empty sign of an ethical awareness of all being, turns this alienation into the "natural" condition of all existence. Whereas posthumanism begins with the premise that, as Nietzsche puts it, "We are from the very beginning illogical and thus unjust beings" for whom a world without inequality is an arrogant fantasy,[19] we argue, following Marx, that humans are *social beings*: they are a product of their collective labor in all forms—both manual and intellectual—that enables them to transform both themselves and their conditions of existence. The advances of industry, science, culture, and technology are the expressions of the *social life* of humanity as it develops the ability to consciously and collectively transform nature and, in the process, transform itself as "[humanity] is a part of nature."[20] The history of humanity—the relations between humans, nature, and things—is therefore the history of humans' modes of production. In this sense, capitalism is an expression of the social life of humanity, but it is an expression of social life divided against itself that through the mechanism of exploitation turns the "life activity" of humanity into "a mere means to [its] *existence*."[21] Under capitalism, a system that takes the collective labor of humanity and directs it toward the generation of wealth for the few, the means by which humanity exists as a species becomes the means by which each individual is forced to survive in an economic competition that divides people from each other and from nature. Through private ownership of the means of production, capitalism strips from the majority of humanity everything needed to live except their labor-power, and thus turns the "species life" of humanity in *the collectivity of labor* into the "natural life" of class society and the alienation of individual *wage-labor*. The species-being of humanity, the "free, conscious activity" of collective labor,[22] is denied by capital's ideologists in order to propagate the idea that there are only individuals as they exist within class society: no collectivity of labor and therefore no alternative to the existing.

Insofar as the concept of the "natural" remains suspect in the annals of contemporary cultural theory, the reduction of class life to natural life takes the form today of a postnatural, postsocial and, above all, posthistorical *ontology* in which class relations disappear and (bourgeois) individuality is (again) reborn as the difference constitutive of *posthuman* "common being." In the words of Roberto Esposito, common being "is the interval of difference, the spacing that brings us into relation with others in a common non-belonging, in this loss of what is proper that never adds up to a common 'good.'"[23] Truly enlightened, in this narrative, are those today who recognize humans' fundamental ontological state of "non-belonging." The implication here is that the historical and material contradictions of social life that prevent realizing a just society are not the problem. It is instead humans' reliance on ("totalizing") notions (e.g., "common good"), since these suppress the difference of the (non)human by falsely assuming human ideas can ever represent, much less understand, the world. More specifically, according to the arguments for the need for a "posthumanism" turn, the various social and cultural failures of capitalism, whether in regards to meeting people's needs or protecting the environment, are not a consequence of production for profit, but the "correlationism" at the heart of Western philosophy; namely, "the idea according to which we only ever have access to the correlation between thinking and being, and never to either term considered apart from one another."[24] In this Nietzschean narrative, the imposition of "human knowledge" on being results in the erasure of individuality, uniqueness, and difference, as well as the failure to recognize the nonhuman agents that shape and reshape reality on their own terms indifferent to human interests.

Posthumanism displaces the praxis of critique—the explanatory critique of world-historical material relations of exploitation that are bringing about contemporary crises and new struggles toward the collective transformation of these relations—on the grounds that critique requires the "humanist" presumption of "knowing" and thus reproduces the "correlationism" that "habitually [grasps] the world in terms of our own preimposed concepts."[25] Instead, the dominant posthumanism contends that what we are witnessing is not simply an environmental or social crisis which can be resolved through human intervention, but an epistemological failure to realize the extent to which the world and all manner of nonhuman actants operate alongside of and independently of human intentionality or understanding.[26] As Thacker argues, "we cannot help but to think of the world as a human world, by virtue of the fact that it is we human beings that think it" but "one of the greatest challenges that philosophy faces today lies in comprehending the world in which we live as both a human *and* non-human world."[27] As a result, at a time when human intentionality is said to be the cause of so many "large" (and self-inflicted) crises—"planetary disasters, emerging pandemics, tectonic shifts, strange weather, oil-drenched seascapes, and the furtive, always-looming threat of extinction"[28] —the world as conceived of by humans as "for us" is therefore "increasingly unthinkable"[29] and all forms of "human-centric" thought, including the praxis of critique, are declared to be "inadequate" in the era of "global climate change."[30] What is needed instead, according to posthumanists, is an ethical turn to a conception of the "world-in-itself-for-us"; that is, the world as a "hidden" realm that is "indifferent to human knowledge, much less to our all-too-human wants and desires."[31] The posthumanism turn thus predicates the collective project of securing the social good on "acknowledging" the "nonhuman [. . .] as a being in its own right;"[32] that present in "all beings whatsoever: eraser, black hole singularity, ceramic knife, molasses, slug" is an "infinite feeling of inner space" that makes them the "spiritual" equals to humans.[33]

Because posthumanism has become the dominant cultural logic and its terms and assumptions now shape the daily common sense, our critique may be dismissed as repressive of the "real"-ity of posthumanism and *élan vital*. Critique as a social praxis is made out to be an "outdated" practice in the time of posthumanism because of what is said to be its "speciesist" assumption that life can be grasped and transformed through the movement of "reason." However, critique is not the movement of pure reason as it is represented in humanist discourses, nor is it the figure of undecidability immanent in textuality that undoes the metaphysics of humanism, as it is in the post-humanist philosophy that traces itself through the writings of Heidegger, Derrida, and deMan. Both these understandings of critique assume that critique is without material foundation and has no basis in reality. The basis of critique, however, is, as Marx says, "man himself,"[34] man, that is, not as "an abstract being squatting outside the world," but men "as they actually are; i.e. as they act, produce materially, and hence as they work under definite material limits, presuppositions and conditions independent of their will."[35] In other words, the critique of humanism is not a rejection of all that makes man appear "other than himself," as Marx puts it in his early writings, but a situating of the human within what Marx calls "species being."[36] This entails getting at the root of what constitutes our being-in-common while paying close attention to the ways in which human productions negate themselves and estrange the producers from their collectivity and nature. To be clear: Marx is the original thinker of the "(post)human" in terms of thinking the human not as an ahistorical essence but in terms of "the ensemble of the social relations."[37] In Marx's writings, however, the material conditions through which human beings produce themselves are not reified as an occult "other" realm, but instead grasped historically as an ongoing process of the transformation of the natural-social world by collective labor. The (post)humanism in Marxist thought, in other words, is *not* the merely speculative "posthumanism" of the contemporary but the materialism of the social relations which historically produce the human (differently) as human. Marxist (post)humanism, in other words, is a materialist inquiry into the class conditions of the estrangement of species-being in production for profit so as to transform capitalism. It is just such inquiry that is blocked by the dominant posthumanist assumption that rejects any "figure which makes man into a species"[38] on the grounds that to do so reduces politics to the biopolitics of "state philosophy," what Derrida calls the regime of "carnophallogocentrism," with its "interventionist violence that is practiced [. . .] in the service of or for the protection of the animal, but most often the human animal."[39] Rather, posthumanist theory purports to define "the nature of reality independently of thought and of humanity more generally,"[40] and, by avoiding the projection of "an epistemological problem into the ontological realm," to prevent "the basic determination of 'what is'" from being "a contentious political matter."[41] Indeed it is just this erasure of species-being as the ontological horizon of the human—both its foundation and, in class societies, its negated "essence"—which explains why posthumanism now has become the dominant mode of intelligibility that naturalizes and justifies the contradictions of the capitalist mode of production and, especially, the ruthless binary of class.

Posthumanism, we argue, borrows a move of the old Young Hegelians by taking the historical relativity and incompleteness of human knowledge about the world as the decentering of thought within itself and assumes that the self-alienation of thought is a sign of the existence of a spiritual common of nonhuman others that humanity must learn to live with for the good of all. "Ontology" becomes the sign of a self-destructing kernel that exposes the limits of human epistemologies and calls our attention to the inherent unknowability of being itself. Robert Pepperell thus argues that the "technological advances [. . .] in the fields of micro-electronics, gene manipulation, and communications" that "point to the possibility of controlling, synthesising or even surpassing aspects of nature which until now have eluded our command,"[42] have ironically revealed that "our capacity to order and control the universe is ultimately limited,"[43] thus ushering us into a posthuman condition in which "we no longer find it necessary, or possible, to distinguish between humans and nature."[44] On this view, the incompleteness of our scientific control over nature is taken to mark an absolute categorical limit that militates against the capacity of humans to produce positive and reliable knowledge of the nonhuman world such as is required to wrest a "realm of freedom" from the "realm of necessity."[45] By taking the historical relativity of knowledge and making it into an absolute principle about the "unknowable" nature of reality, posthumanism asserts that because the material world that exists outside our heads is not always or absolutely conceptually grasped there is something "wholly other" about it such that it can be neither known nor transformed. Ontology, in this context, comes to represent the "hidden" nonlocalizable materiality that cannot become an object "for us," like Kant's notion of the "noumenal" realm that exists "in itself" independently of the "phenomenal." The radical alterity of this "other" world does not prevent posthumanism from finding there, however, "enmeshed" networks of human and nonhuman "assemblages" that, they insist, announce a cognitive break from Enlightenment anthropocentrism: the arrival of a planetary consciousness for which binary relations are outmoded mental impositions on the world, not the effect of material social structures. The posthumanism turn thus turns theory into a local, speculative, and contingent meditation with no knowable connections to the contemporary "crises," ecological and otherwise, that are said to be the central concern of the posthumanities, and thereby makes the world safe for capital by immunizing it from critique. The ideological effects of posthumanism's accomodationism are, in this context, not avoided by the emerging theoretical movement known as "accelerationism," which instead of rejecting humanist reason and techne, claims to "side with the emancipatory dynamic [of capitalism] that broke the chains of feudalism" by embracing technology and ostensibly seeks to accelerate the end of the property relations that fetter technological and scientific advance.[46] While the accelerationists have been depicted as "transhumanist" in their "Promethean" embrace of technology and human progress,[47] they in fact update posthumanism's occulting of class in a more radical guise. Responding to the immanent failures of posthumanist atomizing of the social and dismantling of any possibility of social transformation, they call for "organizing a broadly populist left, building the organizational ecosystem necessary for a full-spectrum politics on multiple fronts, and leveraging key points of power whenever possible."[48] They nonetheless continue the North Atlantic left's dismissal of class exploitation as the basis of capitalism. Proposing that a twenty-first century post-capitalist movement has to revise Marxism's class critique beyond recognition on the grounds that "this analysis failed to attend to both the range of possible liberating struggles (based in gender, race, or sexuality) and the ability of capitalism to restructure itself,"[49] the accelerationists' call for a post-work society that "remains irreducible" to class interests[50] simply, and one-sidedly, inverts other posthumanists' denunciation of technology and promotes "cryptocurrencies" and "social networks" as the starting points for a new public sphere. Like other iterations of posthumanism, accelerationism occults the class basis of the social as "community."

Human, All Too (Post)Human is a red critique of the posthumanism turn which, as in Marx's inversion of Hegel, "turns it right side up again" by seeking the causes of capitalism's crises not in the "general consciousness" but in "the real community."[51] Posthumanism, we argue, dematerializes both humanism and posthumanism. It represents the class relations of capitalist society and their proliferating consequences on human and nonhuman life as an effect of how we think. We argue, instead, that how we think about the human is shaped by the labor relations that in complex and sometimes highly mediated ways shape all aspects of social and cultural, human and nonhuman life. Humanism is not primarily a self-centered worldview; it is a ruling-class ideology. It reflects a historical arrangement of human relations, one in which the social priority is not meeting common needs but the private accumulation of wealth. These class relations have shaped the way humans and nature are treated: not as ends in themselves, but as means for expanding the accumulation of profit. Posthumanism, however, removes class relations from the landscape of humanism and posthumanism and thereby underwrites the authority of capital over the social. Each of the texts included in this collection is, instead, guided by the praxis of critique, which begins by regrounding "*general* consciousness [as] only the *theoretical* shape of that of which the *living* shape is the *real* community"[52] in order to "show the world what it is really fighting for," as such "consciousness is something that it has to acquire, even if it does not want to."[53] In contrast to the posthumanist imaginary, *Human, All Too (Post)Human* advances the critique that what is presented as an ethical turn toward the nonhuman is an extension of rather than an opposition to the idealism and human-centeredness of humanism; in fact, it radicalizes the reformism of earlier humanist projects for change by essentially declaring the extinction of the great collective struggles for social emancipation, and naturalizing humanity under capitalism as humanity as such. Through an analysis of the diverse discourses of the posthumanism turn—from "animal studies" and "common-ism" to "object-oriented-ontology," "biopolitics," and "accelerationism"—we demonstrate why the logic of posthumanism is a ruling class ideology that in the name of "entanglements" *dis*-entangles the effects of capitalism from the cause of contemporary crises in capitalist production and presents the image of a world beyond understanding or transformation. We argue that it is only the negation of capitalism and its private ownership of the means of production—namely, the full realization of the collectivity of labor that is communism—that represents the "*genuine* resolution of the conflict between man and nature and between man and man—the true resolution of the strife between existence and essence, between objectification and self-confirmation, between freedom and necessity, between the individual and the species."[54]

BIBLIOGRAPHY

Badiou, Alain. *The Century*. Translated by Alberto Toscano. Cambridge, UK: Polity Press, 2007.

Braidotti, Rosi. *The Posthuman*. Cambridge, UK: Polity Press, 2013.

Bryant, Levi, Nick Srnicek, and Graham Harman. "Toward a Speculative Philosophy." In *The Speculative Turn: Continental Materialism and Realism*, edited by Levi Bryant, Nick Srnicek, and Graham Harman. Melbourne, Australia: re.press, 2011.

DeLanda, Manuel. *Intensive Science and Virtual Philosophy*. New York: Bloomsbury, 2013.

Derrida, Jacques. *The Animal That Therefore I Am*. Translated by David Wills. New York: Fordham University Press, 2008.

---. *The Gift of Death*. Translated by David Wills. Chicago, IL: University of Chicago Press, 1995.

---. *Without Alibi*. Edited and Translated Peggy Kamuf. Stanford, CA: Stanford University Press, 2002.

Engels, Frederick. "The Part Played by Labour in the Transition from Ape to Man." In *Marx-Engels Collected Works*, vol. 25. New York: International Publishers, 1975.

Esposito, Roberto. *Bios: Biopolitics and Philosophy*. Translated Timothy Campbell. Minneapolis: University of Minnesota Press, 2008.

---. *Communitas: The Origin and Destiny of Community*. Translated Timothy Campbell. Stanford, CA: Stanford University Press, 2010.

Grusin, Richard A. *The Nonhuman Turn*. Minneapolis: University of Minnesota Press, 2014.

Haraway, Donna. *When Species Meet*. Minneapolis: University of Minnesota Press, 2008.

Latour, Bruno. *The Politics of Nature: How to Bring the Sciences into Democracy*. Translated by Catherine Porter. Cambridge, MA: Harvard University Press, 2004.

---. *Reassembling the Social: An Introduction to Actor-Network Theory*. Oxford: Oxford University Press, 2005.

Mackay, Robin and Armen Avanessian. Introduction to *The Accelerate Reader*, edited by Robin Mackay and Armen Avanessian, 1–46. Falmouth, UK: Urbanomic Media, 2014.

Marx, Karl. *Capital Volume 3*. In *Marx-Engels Collected Works*, vol. 37. New York: International Publishers, 1998.

---. *Economic and Philosophic Manuscripts of 1844*. In *Marx-Engels Collected Works*, vol. 3. New York: International Publishers, 1975.

---. Introduction to *Critique of Hegel's Philosophy of Right*. In *Marx-Engels Collected Works*, vol. 3. New York: International Publishers, 1975.

---. "Letter to Ruge." In *Marx-Engels Collected Works*, vol. 3. New York: International Publishers, 1975.

Marx, Karl and Frederick Engels. *The German Ideology*. In *Marx-Engels Collected Works*, vol. 5. New York: International Publishers, 1976.

Meillassoux, Quentin. *Against Finitude: An Essay on the Necessity of Contingency*. Translated by Ray Brassier. London: Continuum, 2008.

Morton, Timothy. *Hyperobjects: Philosophy and Ecology after the End of the World*. Minneapolis: University of Minnesota Press, 2013.

Nietzsche, Friedrich. *Human, All Too Human: A Book for Free Spirits*. Translated by R. J. Hollingdale. New York: Cambridge University Press, 1986.

Pepperell, Robert. *The Posthuman Condition: Consciousness Beyond the Brain*. Bristol, UK: Intellect Books, 2003.

Roden, David. "Accelerationism and Posthumanism II." *enemyindustry: philosophy at the edge of the human*. (blog). November 21, 2013. http://enemyindustry.net/blog/?p=4531.

Shaviro, Steven. *The Universe of Things: On Speculative Realism*. Minneapolis: University of Minnesota Press, 2014.

Srnicek, Nick and Alex Williams. *Inventing the Future: Postcapitalism and a World Without Work*. London: Verso Books, 2015.

Stiegler, Bernard. *Technics and Time 1*. Translated by Richard Beardsworth and George Collins. Stanford, CA: Stanford University Press, 1998.

Thacker, Eugene. *In the Dust of this Planet*. Winchester, UK: Zero Books, 2011.

NOTES

1. Richard A. Grusin, *The Nonhuman Turn* (Minneapolis: University of Minnesota Press, 2014), vii.
2. Jacques Derrida, *The Animal That Therefore I Am*, trans. David Wills (New York: Fordham University Press, 2008), 41.
3. Donna Haraway, *When Species Meet* (Minneapolis: University of Minnesota Press, 2008), 46.
4. Bernard Stiegler, *Technics and Time 1*, trans. Richard Beardsworth and George Collins (Stanford, CA: Stanford University Press, 1998), 26.
5. Manuel DeLanda, *Intensive Science and Virtual Philosophy* (New York: Bloomsbury, 2013), 51.
6. Bruno Latour, *Reassembling the Social: An Introduction to Actor-Network Theory* (Oxford: Oxford University Press, 2005), 50-52.
7. Timothy Morton, *Hyperobjects: Philosophy and Ecology after the End of the World* (Minneapolis: University of Minnesota Press, 2013), 7.
8. Rosi Braidotti, *The Posthuman* (Cambridge, UK: Polity Press, 2013), 136.
9. Roberto Esposito, *Communitas: The Origin and Destiny of Community*, trans. Timothy Campbell (Stanford, CA: Stanford University Press, 2010), 139.
10. Karl Marx, introduction to *Critique of Hegel's Philosophy of Law*, in *Marx-Engels Collected Works*, vol. 3 (New York: International Publishers, 1975), 175.
11. Friedrich Nietzsche, *Human, All Too Human: A Book for Free Spirits*, trans. R. J. Hollingdale (New York: Cambridge University Press, 1986), 35.
12. Karl Marx, *Economic and Philosophic Manuscripts*, 306.
13. Roberto Esposito, *Bios: Biopolitics and Philosophy*, trans. Timothy Campbell (Minneapolis: University of Minnesota Press, 2008), 110–45.
14. Jacques Derrida, *Without Alibi*, ed. and trans. Peggy Kamuf (Stanford, CA: Stanford University Press, 2002), 75–76.
15. Jacques Derrida, *The Gift of Death*, trans. David Wills (Chicago, IL: University of Chicago Press, 1995), 71.
16. Eugene Thacker, *In the Dust of this Planet* (Winchester, UK: Zero Books, 2011), 52.
17. Ibid., 53.
18. Ibid., 6.
19. Friedrich Nietzsche, *Human, All Too Human*, 28.
20. Karl Marx, *Economic and Philosophic Manuscripts*, 276.
21. Ibid., 276.
22. Ibid., 276.
23. Roberto Esposito, *Communitas*, 139.
24. Quentin Meillassoux, *Against Finitude: An Essay on the Necessity of Contingency*, trans. Ray Brassier (London, UK: Continuum, 2008), 5.
25. Steven Shaviro, *The Universe of Things: On Speculative Realism* (Minneapolis: University of Minnesota Press, 2014), 66.
26. Bruno Latour, *The Politics of Nature: How to Bring the Sciences into Democracy*, trans. Catherine Porter (Cambridge, MA: Harvard University Press, 2004), 75.
27. Eugene Thacker, *In the Dust of this Planet*, 2.
28. Ibid., 1.
29. Ibid., 1.
30. Ibid., 4.
31. Ibid., 53–54.

32. Timothy Morton, *Hyperobjects*, 119.
33. Ibid., 120.
34. Karl Marx, introduction to *Critique of Hegel's Philosophy of Law*, 182.
35. Karl Marx and Frederick Engels, *The German Ideology*, in *Marx-Engels Collected Works*, vol. 5 (New York: International Publishers, 1976), 35–36.
36. Karl Marx, *Economic and Philosophic Manuscripts*, 275.
37. Karl Marx, "The Theses on Feuerbach," in *Marx-Engels Collected Works*, vol. 5 (New York: International Publishers, 1976), 4.
38. Alain Badiou, *The Century*, trans. Alberto Toscano (Cambridge, UK: Polity Press, 2007), 174.
39. Jacques Derrida, *The Animal That Therefore I Am*, 25.
40. Levi Bryant, Nick Srnicek, and Graham Harman, "Toward a Speculative Philosophy," *The Speculative Turn: Continental Materialism and Realism*, ed. Levi Bryant, Nick Srnicek, and Graham Harman (Melbourne, Australia: re.press, 2011), 3.
41. Ibid., 16.
42. Robert Pepperell, *The Posthuman Condition: Consciousness Beyond the Brain* (Bristol, UK: Intellect Books, 2003), 161.
43. Ibid., 167.
44. Ibid., 161.
45. Karl Marx, *Capital Volume 3*, in *Marx-Engels Collected Works*, vol. 37 (New York: International Publishers, 1998), 807.
46. Robin Mackay and Armen Avanessian, introduction to *The Accelerate Reader*, ed. Robin Mackay and Armen Avanessian (Falmouth, UK: Urbanomic Media, 2014), 7.
47. David Roden, "Accelerationism and Posthumanism II," *enemyindustry: philosophy at the edge of the human* (blog), November 21, 2013, http://enemyindustry.net/blog/?p=4531.
48. Ibid., 175.
49. Ibid., 14.
50. Ibid., 174.
51. Karl Marx, *Economic and Philosophic Manuscripts*, 298.
52. Ibid., 298.
53. Karl Marx, "Letter to Ruge," in *Marx-Engels Collected Works*, vol. 3 (New York: International Publishers, 1975), 144.
54. Karl Marx, *Economic and Philosophic Manuscripts*, 296.

I

"Natural" Life and "Species" Life

ONE

The New Class Common-Sense

Biopolitics, Posthumanism, and Love

Jennifer Cotter

As class contradictions have sharpened with the intensification of the crisis of profitability in global capitalism, cultural theory in the North Atlantic retreats further and further into *spiritual explanations and resolutions* of material contradictions. Rather than confronting the relationship of the poverty of workers around the world, the concentration of wealth into fewer hands, increases in unemployment and economic insecurity, the commodification of all aspects of life subordinating them to production for profit, and the deepening alienation of workers *to class relations founded on exploitation in production*, contemporary cultural theory has abandoned examining the *social, the historical, and the material* basis of these questions in production relations. Instead, it is turning to immaterial, affective, and spiritual explanations and resolutions of material contradictions and ideologically translating capitalism and exploitation into ontological conditions of life as such.

The Biopolitical Turn, or, Disappearing Class from Life under Capitalism

It is in the aforementioned context that cultural theory in the global North has embraced what Christian Marazzi calls "the biopolitical turn of the economy"—an increasing turn to "biopolitics" as a means to explain and address the social and economic contradictions in capitalism now.[1] At the core of biopolitical theory is a substitution of "life"—particularly the spiritualist concept of a creative "life-force" or what Henri Bergson calls *élan vital*—for the historical and material relations of the *dialectical praxis of labor* and *class* as explanations of the material basis of contradictions in capitalism now and their transformation. In other words, central to biopolitics is a cultural spiritualism which ideologically translates historical and material relations into a transhistorical and autonomous power of life—a mystical vitalism—and posits this spiritual vitalism as the basis of bringing about new social relations. This new "spiritualism" of life is offered in a variety of articulations in contemporary cultural theory, from writings that overtly address theories of biopower and biopolitics as in the writings of Michael Hardt and Antonio Negri (*Commonwealth*), Giorgio Agamben (*Homo Sacer*), Maurizio Lazzarato, and Antonella Corsani, among others, to the transspecies posthumanist writings of Derrida (*The Animal That Therefore I Am*), Agamben (*The Open*), and Donna Haraway (*When Species Meet*). These biopolitical theories claim to address a range of material problems that are the effect of production for profit in capitalism: from the encroachment of commodity relations into all aspects of life, including the private ownership of strands of DNA and whole species of plants and animals; to the degradation of the environment in the interests of profit; to economic crisis; to the extension of the working day; to the subordination of love and sexuality to production for profit; to the estrangement of workers from social wealth. . . Yet, at the same time, biopolitical theories abstract these problems from their origins in class relations and exploitation. In place of addressing the material conditions and relations that have given rise to these problems, biopolitical theories posit an "other-world" and an "other-worldly life"—a concept of *spiritual* life that is prior to, constitutive of, transcendent of and/or outside of the historical and social relations of capitalism—as the basis of a new "common" beyond the material contradictions of capitalism.

The spiritualization of life in biopolitical theories, to be clear, is represented as a new form of materialism. The substitution of "life" for "class" in biopolitical theories draws from Foucault's theory of "bio-power" in which he argues that "modern man is an animal whose politics calls his existence as a living being into question."[2] "Bio-power," Foucault contends, is a regime of power that, rather than ruling by threat of death, produces life through the disciplining of bodies, the regulation of populations, and through the "technologies of the self" in which bodies come to bind themselves to identities produced through sovereign power. In fact, Foucault posits "bio-power"—the instrumental disciplining of bodies such that they come to experience their own subjection as the norm of life and source of pleasure—rather than the exploitation of labor, as the material basis of capitalism. Capitalism, Foucault contends, is not possible without "bio-power": "bio-power was without question an indispensable element in the development of capitalism; the latter would not have been possible without the controlled insertion of bodies into the machinery of production."[3] According to Foucault, "There is no binary and all encompassing opposition between rulers and ruled at the root of power relations."[4] In this view there is "no regulative mechanism" of power relations. Instead, in Foucault's cultural imaginary of power, "power is everywhere . . . because it comes from everywhere."[5]

In this narrative, an explanatory critique of power as the effect of class relations of capitalism—relations between exploiter and exploited—is regarded as part of a binary metaphysics of power which discursively imposes a regime of truth (power/knowledge) onto what is actually, so the story goes, an ineffably and mysteriously plural, diffuse, and amorphous multiplicity of forces. On the one hand, life is assumed to spontaneously produce a discursive proliferation of meanings (knowledges) that then discipline and contain it. On the other hand, life is regarded as an ineffable and plural opacity that resists all conceptual explanation. The subjection of bodies is reduced to a contingent, "spontaneous" and aleatory effect of "life" *as such* or the sheer fact of living. This makes it appear as if "power over life" comes not from structural relations of exploitation, but from a non-structural, amorphous, cultural plurality—a cultural democracy of power from below to which all have access by virtue of living—that is "everywhere."

Foucault's theory of bio-power is not a form of materialism but a form of cultural spiritualism. When Foucault argues that "bio-politics" is at the root of capitalism, he dehistoricizes "the machinery of production" into which he claims bodies are "inserted." The existence of "the machinery of production"—or a "controlled insertion" of bodies—is itself the effect of the dialectical praxis of labor. This is because power is not an autonomous, trans-historical life force nor is it an ineffable diffuse plurality beyond historical and conceptual explanation, but an effect of definite historical and material conditions and relations. Power, in other words, rests upon material conditions of production. Whether or not the society has the power to end starvation or to condemn the majority of the laboring population to a lifetime of starvation, has to do with the level of development of its material conditions of production—its forces of production—and the social relations of production (the labor and property relations) that determine the social ends and interests toward which labor is put. This is another way of saying that power is the historical and material effect of labor in the form of property. In a society in which property is privately owned, power is the capacity of the ruling class to "command over the surplus-labor" of workers in production.[6] At the root of power relations is an antagonistic class relation: the antagonism between owners of the means of production and workers who only own their labor to sell in order to survive and are exploited. The binary of class is historical and material, not, at root, discursive: class binaries are not the effect of nature, God, nor are they the effect of "Western metaphysics," "discursive construction," "binary thinking," or conceptualization, but the effect of private ownership of the means of production.

Foucault's theory of power does the ideological work of capital by concealing and inverting the structural relations of class in capitalism. In place of the material transformation of structural relations of capitalism, Foucault advocates "resistance" within—a change in the discursive and cultural regimes and a re-valuing of "life"—as the basis of a "different economy of bodies and pleasures."[7] This amounts to the updating of the culture of capitalism as the limit of change while the needs of the masses for material abolition of exploitation is dismissed as a reactionary nostalgia for the impossible—what Foucault dismisses as "The 'right' to life . . . beyond all the oppressions."[8] Changing the cultural values of life and regarding this as *constituting* material change—i.e., as an end in itself—becomes a means to ideologically update power relations without fundamentally transforming them.

The spiritualism that is implicit in Foucault's theory of "biopower" has become more pronounced in contemporary articulations of biopolitics. In their most recent book, *Commonwealth*, for example, Michael Hardt and Antonio Negri put forward the concept of "biopolitics" as a supplement to Foucault's "biopower."[9] While "biopower" they argue, is the power "over life," they deploy the concept of biopolitics to argue for an autonomous "power of life" to resist.[10] In doing so, they posit an abstract and essentially spiritualist concept of a creative life force—a neo-Augustinian conception of life—outside the social and historical as the basis of a new commonwealth and "mass exodus" from capitalism. Giorgio Agamben, *in contrast* to Hardt and Negri, uses the term *biopolitics* instead to refer to the "mechanisms and calculations of State power" over life—what Foucault calls *biopower*—rather than an autonomous "power of life."[11] For Agamben, what Hardt and Negri call the "power of life" to resist, is always already an extension of the sovereign "power over life" not a resistance to it. The "originary moment" of politics, according to Agamben, is that natural life, "the simple fact of living" or "life common to all" (what the ancient Greeks called *zoē*) has been subsumed, "captured" by "biopolitical regimes of power" so that precisely when natural life (*zoē*) is posited to be "outside" of sovereign power (*bios*)—banned by it, excluded from it—this is actually an extension of the inside of sovereign power.[12] Politics, in other words, is always already the inclusion of this "excluded" life, the politicizing of natural life—the reduction of bodies to de-sacralized or "bare life."[13] Yet, despite their different theories and different semantic uses of the term biopolitics, Agamben's biopolitical theory as a whole is predicated—as much as is Hardt and Negri's theory—on restoring a spiritualist concept of life. In contrast to the life of "*bios* and *zoē*"—at the interstices of which, Agamben contends, is always already de-sacralized "bare life"—Agamben posits a "new form-of-life," a "messianic redemption" or "happy life" as a moment of transcendence.[14] Finding even Foucault to be too historical for positing biopower as a stage in the development of history, and history itself to be always already a reduction of life to bare life, Agamben puts forward this new form of life as outside of history. For Agamben, there is no *historical* possibility of ending exploitative social relations and bringing about freedom through material transformation; *all* social, political, and historical life is always already a violent subordination of life to biopolitics. All historical and social life, regardless of the actual social and economic organization of society, is in this view always already a form of "desacralized" life or "bare life."[15] Rather than articulating a materialist theory of social transformation and the historical and material possibility of bringing about social relations free from exploitation, Agamben articulates a spiritual song of mourning and melancholia for a "lost" sacred life that has never been: "Redemption is not an event in which what was profane becomes sacred and what was lost is found again. Redemption is, on the contrary, the irreparable loss of the lost."[16] Social transformation is reduced to a spiritual journey toward the "messianic" common life "to come" that can never fully be materialized without becoming "bare life"—a journey, in other words, in which the working class, like Ralph Ellison's invisible man at the Battle Royal, is "kept running."

This essentially spiritualist understanding of life, moreover, is codified on a new level within a specific variant of biopolitical theories—transspecies posthumanism—which posits a common spiritual life beyond historical differences between "humans" and "animals" as the basis of new global social relations. As a form of biopolitical ideology, transspecies posthumanism displaces class relations with an ahistorical and spiritualist understanding of common life. Transspecies posthumanism does so more specifically by declaring the historical differences between species—specifically between humans and other animals—as metaphysical abstractions of the common fact of living that exists within all species. One of the main goals of transspecies posthumanism is to divert attention away from class relations and exploitation of surplus-labor, by enacting a "fissure" in the concept of the human—that is, by ideologically dissolving the historical difference between human and animal—and, in so doing, invoking a "crisis" in the concept of human labor-power. Transspecies posthumanism, therefore, situates "life"—which it understands as "transspecies" life or life common to all species—outside of the historical relations that produce differences between the species. In doing so it dispenses with projects for material transformation of historical relations of production on the grounds that they are violently anthropocentric and function on the basis of what Donna Haraway calls "the goad of human exceptionalism."[17]

One of the central concepts through which biopolitical and transspecies posthumanist theories advance a new spiritualism is in their theorizations of "love." "Love" is re-articulated in these discourses as an autonomous life-force that will bring into being new social forms and particularly a new "true," alternative globalization: what Hardt and Negri call an "ontological event" which brings into being the "common," and what Donna Haraway calls an "other"-world but is, in actuality, a spiritualist theory of the "other-worldly" which provides an ideological space for the privileged to accommodate capitalism and the exploitation of labor of the majority, in the name of a "resistant" and alternative globalization. In particular, post-nuclear and transspecies love, family and/or sexual relations are put forward in biopolitical and posthumanist theories as "constitutive" of a new world order. This essay addresses the class politics of the new biopolitical and transspecies posthumanist spiritualism and especially what they offer as "resistant" theories of questions of life, love, family, and sexuality now. In particular, this essay critiques the class politics of Hardt and Negri's argument that "love" is a "biopolitical event" that constitutes the common, including their seemingly radical argument in *Commonwealth* for a "mass exodus" from the family and capitalism, which they argue is an institution which corrupts the creative forces of biopolitical labor and prevents the multitude from bringing about commonwealth in society.[18] As well, this essay critiques the class politics of Donna Haraway's transspecies posthumanist theory of love in which intimacy and "love" with companion species are represented as a radical "other-world-making" evolution and capitalism is represented as evolved and the regime of the evolved.

The argument that biopolitics is a form of spiritualism, to be clear, is at odds with the *claims* and self-representations of biopolitical cultural theorists, who contend that their theories are a "new" and "true" form of materialism. Biopolitics maintains that capitalist relations of production have been fundamentally *materially* transformed by the development of bio-technologies, cybertechnologies, knowledge work, the growth of service industries, the erosion of industrial manufacturing in the North. . . so that earlier distinctions between "productive" and "reproductive" labor have collapsed. Antonella Corsani, for example, claims that "what is emerging from the metamorphoses of capitalism is a new relationship between capital and life."[19] "The sphere of reproductive activities," Corsani contends, "is integrated into that of production, so that 'life itself' is productive of surplus-value" whether we are eating, drinking, "even," she claims, "when we are sleeping or making love."[20] Michael Hardt and Antonio Negri confirm this assumption when they suggest that we should no longer speak of capitalism in terms of productive labor but of "biopolitical labor" (a trope for reproductive labor), which produces social life itself or "subjectivities." In *Commonwealth*, Hardt and Negri displace "exploitation" with "alienation" as the key site of struggle and social transformation when they declare: "we find ourselves being pulled back from exploitation to alienation, reversing the trajectory of Marx's thought."[21] According to Hardt and Negri, "alienation" has no material relation to exploitation (the theft of surplus-labor by owners of the means of production during the working day) and this, they claim, is owing "to the fact that some characteristics closely tied to exploitation particularly those designating capital's productive role, have faded."[22] On this basis biopolitical theories posit reproduction—what Hardt and Negri refer to as "biopolitical production"—as having materially displaced production in capitalism.

What has actually been occurring in transnational capitalism, however, is not a disappearance of productive labor or exploitation but the transfer of productive labor and the export of capital from the global North to the global South in search of securing sources of cheaper labor to exploit. As Paula Cerni has argued, "something very material has accompanied the creation of a 'post-material' economy where 83 percent of non-farm employees work in services."[23] Far from actually bringing about a "post-material" economy "the real shift towards [unproductive] service sectors in Western economies" has resulted in a situation in which Western economies "no longer produce enough goods to fund [their] own massive physical requirements, and, as a result, [they are] running an unprecedented trade deficit."[24] What is at the root of this is the fact that it is labor not the "immaterial" of culture or ideology that is the source of social wealth. It is precisely because the basis of profit has been and continues to be the exploitation of productive labor that the wealth of North Atlantic capital—and its share of the profits of the world market—is in decline as it has concentrated investment in reproductive labor within its own respective national borders, and has relied more heavily on productive labor around the world. To conflate the shifts in the way in which North Atlantic capital aims to acquire a larger share of the social wealth in transnational capitalism, with a fundamental change in the relations through which this wealth is actually produced in transnational capitalism, is a parochial analysis of the global economy that erases the continued exploitation of surplus labor of workers around the world in China, in India, in Pakistan, Iraq, Afghanistan. . . . These shifts in production are not a break from the *class* relations of capitalism and the exploitation of workers around the world; they are an intensification of its irreconcilable class contradictions. Moreover, the consequences of these class contradictions and their "solutions" have been devastating for workers both in the global North and the global South, from the spiking of unemployment, to the loss of homes and pensions, to the gutting of public infrastructure for workers and transferring this social wealth to corporations, to increases in suicide rates, depression, anxiety, and pharmaceutical dependency, to "jobless and wageless recovery" which, in actuality, means an increase in the rate of exploitation of workers.

"Love" as/and the Spiritual Aroma of Exploitation

Biopolitics and transspecies posthumanism, in displacing "class" with "life," "production" with "reproduction," "labor" with "love," are affective and ultimately spiritualist understandings of material contradictions that articulate what Marx calls an "inverted world-consciousness."[25] In "A Contribution to a Critique of Hegel's *Philosophy of Right*: Introduction," Marx critiques religion for the way in which it articulates an inverted world consciousness because, on the one hand, it is "an *expression of* and *protest against* real suffering" and, on the other hand, it provides an "illusory happiness" for "real suffering."[26] By "illusory happiness" Marx means that religion provides an illusory resolution of the material contradictions of exploitation in capitalism that cause the "real suffering" of which religion is both an effect and a response. In this way, rather than providing a material solution to problems of social alienation whose origin are in material relations of production, religion ends up providing a "spiritual aroma" for capitalism that helps to ideologically blur material relations of class and culturally adjust exploited workers to ruling class interests.[27] It is on this basis that Marx argues that "The call [to workers] to abandon illusions about their condition is the call to abandon a condition which requires illusions."[28]

Biopolitics and transspecies posthumanism articulate the "spiritual aroma"—the cultural imaginary—of transnational capital now. They do so by putting forward a "common share" in the "immaterial" of a new "global" culture under capitalism in place of transformation of the material relations of production in capitalism and freedom from exploitation. In doing so they serve to naturalize the material relations of exploitation and culturally adjust the contemporary workforces to the needs of capitalism now. In this respect, biopolitical and transspecies posthumanist theories of love are a continuation—in a new historical form—of updating the working class into a new morality. George Sampson, in his 1921 book on British national education, *English for the English,* provides a telling historical example of this practice in his comments on the role of teaching "English" literature and culture to the working-class: "Deny to working-class children any common share in the immaterial, and presently they will grow into the men who demand with menaces a communism of the material."[29] To put this another way, the "common share" in the "immaterial" of "culture" for all was proposed by representatives of ruling class interests, such as Sampson, in order to ideologically smooth over severe material contradictions which were leading British workers to increasingly call into question the basis of ruling class wealth in their own exploitation. More generally, these comments are symptomatic of the fact that it is in the material interests of capital to provide "immaterial" and "spiritual" resolutions to deflect attention away from the economic and at the same time maintain the cultural cohesion of social bonds that are necessitated by social relations of production founded on exploitation.

Biopolitical and transspecies posthumanist theories of "love" are *ideological* and *illusory* articulations of workers' *actual* need to do away with the conditions that "require illusions." They are a form of "inverted world-consciousness" because they obscure the material need of workers to abolish the material relations under which they are exploited—material relations, that is, which lead to sharpening alienation for workers. This is to say that biopolitical and transspecies posthumanist theories of love are ideological and illusory not because exploited workers do not actually have affective needs; workers do, indeed, have affective needs such as needs for love. Rather, they are ideological and illusory—a form of "inverted world-consciousness"—because they present love as a spiritual force that will heal social alienation without the transformation of material relations founded on private property and exploitation that produce alienation in the first place.

In this sense, biopolitical and transspecies posthumanist theories of love not only articulate the cultural imaginary of capital, but in doing so they fulfill a practical need for capital. This is because changes in social reproduction—including love relations—without fundamental changes in the social relations of production (class relations) are the means by which the ruling class has always ideologically updated the contemporary workforces of capitalism with the skills and cultural intelligibilities needed to be more effectively and profitably exploited in new conditions of production. Biopolitics and its variations such as transspecies posthumanism are aimed at updating, reforming, and crisis managing capital and particularly, producing new subjectivities who will adjust to the needs of capital now.

The general return of contemporary cultural theory to focus on questions of "love"—and, in particular, the spiritualization of love as a creative "life-force" in biopolitical and transspecies posthumanist theories—is, as I have marked above, an ideological response to material contradictions in transnational capitalism now. At the same time, however, "love" becomes a question for cultural theory and a site of contestation because "love" is what historical materialist Alexandra Kollontai calls a "profoundly social emotion."[30] Kollontai means by this that love—and specific forms of love—are *produced* under definite social relations of production and at the same time help to *reproduce* these material relations of production.[31] "Love" plays a cultural and reproductive role in strengthening the social bonds that have been made materially necessary to a given society. On this point, Kollontai argues,

> At all stages of human development love has (in different forms, it is true) been an integral part of culture. [. . .] Love is not in the least a "private" matter concerning only the two loving persons: love possesses a uniting element which is valuable to the collective. This is clear from the fact that at all stages of historical development society has established norms defining when and under what conditions love is "legal" (i.e., corresponds to the interests of the given social collective),

> and when and under what conditions love is sinful and criminal (i.e., contradicts the tasks of the given society).[32]

"Love" is the subject of social contestation because of its *historical* role in strengthening the bonds of social relations which have been made historically and materially necessary to a given mode of production. Love is not transhistorical but socially produced under definite conditions. Specific forms of love are produced and become culturally "valuable" within a given set of material relations of necessity for the way in which they help to reproduce the existing social relations of production. In class society, this means that the dominant forms of love become dominant or culturally valuable because they help to maintain the existing class relations and to reproduce the conditions necessary for maintaining class relations.

For instance, Kollontai remarks that, "the ancient world considered friendship and 'loyalty to the grave' to be a civic virtue" whereas "love in the modern sense had no place and hardly attracted the attention either of poets or of writers."[33] This is owing to the fact that in order to reproduce its conditions of production, the ruling classes of the ancient world "recognized only those emotions which drew its fellow-members close together and rendered the emerging social organism more stable."[34] Moreover, as Kollontai contends, under feudalism, where the feudal household is a site of production of social wealth, marriage is contracted according to the material interests of the family and individuals who choose a married partner against these interests are severely criticized.[35] "Love" in marriage is not emphasized. However, what is valued as a "moral virtue" in feudalism is "chivalrous love" *outside of marriage* by a knight of an inaccessible woman of the ruling class—a "lady" of "nobility" such as the wife of a feudal lord or the queen. This re-articulation of platonic or ideal love—expressed not only in poetry but in military acts of "heroism"—is considered valuable under feudalism for the way in which it serves as a means to bind men of the exploited classes to fight on behalf of the material interests of their feudal lords.[36] Lastly, as Kollontai demonstrates, the dominant sexual morality of "possessive individualism" that develops with the emergence of capitalism is not transhistorical but enabled by historical and material relations of production.[37] More specifically, as the extended feudal household is broken up by the emergence of wage-labor/capital relations, the prevailing sexual morality and ideals of love also change. While capitalism, like feudalism, is a form of class society based on private ownership of the means of production and the private appropriation of surplus-labor, in contrast to feudalism, it entails the dominance of commodity production outside the household as the main form of the production of social wealth and converts labor-power from serf labor tied to the household into a "commodity" on the market. Under these new material relations of production, the family is no longer the main site of the production of social wealth but a site, on the one hand, of *reproduction* of the concentration of wealth into fewer hands (for bourgeois families) and, on the other, of the reproduction of exploitable labor-power by working class families. Correspondingly the prevailing sexual morality and ideals of love are changed. As capitalism develops and the family is pared down, nuclear "family values" and possessive individualism become the cultural ideal and prevent workers from seeing their mutual common interests as a class and to emotionally tie them into the new social relations of reproduction and production.

In historical materialist theory, love and new moral or ethical codes of love, however, do not in themselves bring about new social forms. The shifts in new forms of sexual morality and love are made possible by material developments and changes in the social relations of production. And these new forms of love then serve to reproduce the material relations of production. New social relations of production can only be brought about by means of material praxis (revolution). And in turn, new forms of love presuppose the development of new material conditions for their production. Even forms of love which break from private property relations—for instance, what Kollontai calls "red love" or "love-comradeship"—are not autonomous forces but *presuppose* the emergence of new material relations of production: either new material relations that have been brought into being through revolutionary social transformation or are in the course of formation. A break from private property forms of love, in other words, necessitates the abolition of private property in the relations in production.

The re-turn in contemporary cultural theory to focus on questions of "love" and "affect" is an articulation of the fact that love is a *social* emotion and an integral part of culture that is useful for *reproducing* the social relations of production. Love, and different forms of love, therefore becomes a site of conflict and struggle in cultural theory and in daily life precisely because of the relation of love to material relations. This relation continues today: as class contradictions in capitalism have intensified and more family members have been pulled into the wage-work force, capital also puts pressure on the nuclear-family form insofar as it has begun to serve as a barrier for capital to extract more surplus-labor from the existing workforce. As a consequence a "new" flexible, "post-nuclear," and "posthuman"—but *not* post-class—sexual and moral code of love is emerging. The old morality of love is serving as a hindrance in many cases to the intensification of the exploitation of workers' surplus-labor around the world. The new spiritualism of "love"—in both its biopolitical and transspecies posthumanist variations—is at root an ideological purging of "old" moral codes of love and sexuality once useful to the ruling class during an earlier stage in the development of capitalism and bringing about new moral codes of love and sexuality useful for adjusting workers to the intensification of class contradictions in transnational capitalism now. And yet, biopolitical theories of love ideologically invert the relationship of these new "post-nuclear" and "post-human" moral codes of love to the material relations of production and posit new forms of love as themselves *constituting* new material relations in society. In concealing the relationship of love to class relations and diverting attention away from the need to transform the material relations of production, these spiritualist theories of love also conceal the fact that the "new" "post-nuclear" forms of love and sexuality they promote are not a break from capitalist *relations* of production, but an updating of its social relations of reproduction to adjust workers to the intensification of class contradictions now.

In the discourses of biopolitics, love is abstracted from its relation to the material relations of production and grasped primarily as a trans-social affective and spiritual "life force" that "creates" and brings into being new social forms. Love is "spiritualized." It is represented as a "creative life force" that will heal social alienation in capitalism—which has its origin in the material contradictions of production in capitalism—without actually transforming the material relations of production founded on exploitation. For example, in *Commonwealth* Hardt and Negri argue that "love" is a "biopolitical event" that "produces the common."[38] In other words, their claim is that love brings about new social relations that break from private property and bring about a "commonwealth." To understand their claims about "love as a biopolitical event" that brings about a change in material relations, it is first necessary to understand what Hardt and Negri mean by "biopolitics." Drawing on Foucault's theory of biopower, they make a distinction between "biopower" and "biopolitics." Hardt and Negri deploy the concept of "biopower" to refer to the "disciplinary regimes, architectures of power, and the applications of power through distributed and capillary networks" that are the subject of Foucault's investigations and that he argues do not "repress" but "produce" subjectivities.[39] Hardt and Negri point out that "biopower," although considered by Foucault to be productive of subjectivities rather than "repressive" of pre-existing subjectivities, is nonetheless a concept which discusses regimes of "power *over* life."[40] By contrast, Hardt and Negri use another term, "biopolitics," to refer to what they regard as "the *other* to power (or even an *other* power)."[41] They contend that "there is always a minor current that insists on life as resistance, an other power of life that strives toward an alternative existence."[42] In other words, in contrast to "biopower" which is "power *over* life," "biopolitics," in Hardt and Negri's theorization of it, is the "power *of* life" and, more specifically, the "power of life to resist and determine an alternative production of subjectivity" which "not only resists power but also seeks autonomy from it."[43] Hardt and Negri not only understand "biopolitics" as a striving *toward* autonomy, but as having an *autonomous origin* that transcends the historical and material relations of society: "the biopower against which we struggle is not comparable in its nature to the form of power by which we defend and seek our freedom."[44]

"Biopolitics"—as Hardt and Negri understand it as the "power of life to resist"—is at root a theory of "creative life force," or what Spinoza calls *potenza* and Henri Bergson calls *élan vital,* which has its philosophical roots in spiritual creationisms. "Biopolitics" with its reliance on an autonomous "power of life" to "resist" is a spiritualizing of the dialectical praxis of labor and an erasure of the material relations of production. It translates what Marx calls the "dialectical praxis of labor" into spiritualist terms by abstracting "life" from its material conditions of possibility and ideologically converting productive activity or labor—which exists in a necessary relation to the relations of production—into an autonomous "creativity."

The existence of "life," which is to say "the existence of living human individuals," and the "power to resist" presupposes material conditions which can enable and sustain human life. This is the case since men and women "must be in a position to live in order to be able to 'make history'"; they must be in a position to satisfy needs of "eating and drinking [. . .] habitation, clothing and many other things."[45] In order to satisfy needs to sustain human life, the existence of human life is not only dependent on the means of subsistence, but on *labor*. Labor is, as Engels puts it, not only the source of all wealth but "next to nature," he argues, "it is the prime basic condition for all human existence."[46] There is no "human existence" that is prior to labor and labor is itself not outside of history; it is a dialectical and material relation:

> Labour is, first of all, a process between man and nature, a process by which man, through his own actions, mediates, regulates and controls the metabolism between himself and nature. He confronts the materials of nature as a force of nature. He sets in motion the natural forces which belong to his own body, his arms, legs, head and hands, in order to appropriate the materials of nature in a form adapted to his own needs. Through this movement he acts upon external nature and changes it, and in this way he simultaneously changes his own nature.[47]

The existence of human "life" and its course of development never exists independently of the material conditions of production prevailing at the time (the forces of production) and the social relations within which this production takes place (the relations of production or property relations). And these conditions and relations are themselves the product of past labor and, in turn, shape the course of all other aspects of social life. But labor conditions never remain static: as the forces of production develop this results in the production and satisfaction of new needs which come into direct contradiction with the relations of production, requiring transformation in the relations of production. Human existence is not prior to the social "metabolism" between the forces of production and the relations within which this production takes place and are transformed. As Marx and Engels argue,

> The fact is, therefore, that definite individuals who are productively active in a definite way enter into these definite social and political relations [. . .] [T]he social structure and the state are continually evolving out of the life-process of definite individuals, however, of these individuals, not as they may appear in their own or other people's imaginations, but as they *actually* are, i.e., as they act, produce materially, and hence as they work under definite material limits, presuppositions and conditions, independent of their will.[48]

Biopolitics, by abstracting life from the material relations of production and the dialectical praxis of labor, puts forward an understanding of the "power of life" as limitless. In erasing the relation of necessity between "life" and the dialectical praxis of labor, one of the goals of biopolitics and its ideological renewing of spiritual creationism is, as I discuss further below, to update the contemporary workforces of capitalism to increase their productivity (under the banner of the "power of life") without eradicating exploitation in production. Raising productivity without eradicating exploitation means raising the rate of exploitation of workers with the aim of raising the rate of profit for capital.

By theorizing love as a "biopolitical event," Hardt and Negri understand love as a trans-material, trans-social, and trans-historical "creative life force." At the core of this view is a re-writing of the "social" as ontological. They argue that love, as a biopolitical event, is "productive" by which they mean that it produces "new subjectivities" and "singularities" and, they add, it is productive of new social forms and relations: "When we engage in the production of subjectivity that is love, we are not merely creating new objects or even new subjects in the world. Instead we are producing a new world, a new social life."[49] What Hardt and Negri call relations of "production" are actually relations of *reproduction*. In their theory, the relations of reproduction in capitalism—and specifically the ideological reproduction of subjectivities and the relations through which this takes place—are considered to be the material terrain of social transformation and freedom. The actual material relations of production—the labor and property relations by which social wealth is produced—and the relationship of love to these relations of production are inverted and hidden from view.

The assumption of this theory is that capitalism is held in place at root not by its material relations of production but by the reproduction of subjectivities who will adjust to and participate in capitalist production. Moreover, by this logic, change in subjectivities—or subjective change—is taken to constitute material change as an end in itself. In this narrative subjective change and more specifically, the mere act of "loving"—or "loving differently"—is regarded as an end in itself that "constitutes" new social relations: "love is a process of production of the common and production of subjectivity. This process is not merely a *means* to producing material goods and other necessities but also an end in itself."[50] No transformation outside of "loving differently" and transforming the forms of affective relations is, according to this theory, necessary in order to bring about human freedom. In this theory, not only are the relations of production ideologically "dematerialized" but so are the relations of reproduction themselves. When Hardt and Negri appear to make references to "love" and "love relations" as "social," they do not mean by this that love—and the form that it takes—presupposes specific social relations and material conditions of possibility, rather they theorize "love" as a *transsocial and immaterial* life force that *brings into being* new social forms. Love, they claim, is at root an "ontological event" that constitutes being and reality as such:

> Every act of love, one might say, is an ontological event in that it marks a rupture with existing being and creates new being [. . .] Being, after all, is just another way of saying what is ineluctably common, what refuses to be privatized or enclosed and remains constantly open to all. (There is no such thing as a private ontology.) To say that love is ontologically constitutive then, simply means that it produces the common.[51]

Love, in this argument, is not only outside of social relations but "creates" them. Social relations in other words presuppose love. Love, in this narrative, is both origin (*arche*) and end (*telos*) of reality, or Being as such. At different points in their narrative Hardt and Negri refer to love as: an "economic power," an affective network, a biopolitical event, a biopolitical force which creates and brings into being new social forms, a force which composes singularities (differences) within the common, and the basis of ontology or Being *as such*.[52] In short, in *Commonwealth,* love is understood in theological and spiritualist terms: it is regarded to be all powerful, all encompassing, absolute reality that creates all that is and all that will ever be. It is ideologically "cleansed" from its actual relation to the material relations of production.

To complete their theological sermon on the "power of love," Hardt and Negri argue that love is a "force to combat evil."[53] Here Hardt and Negri deploy the theological concept of "evil" in place of a rigorous historical materialist analysis of material contradictions. Having ideologically displaced the dialectical praxis of labor as the basis of social forms with "love" as a creative life force, the deployment of the concept of "evil" now enables Hardt and Negri to displace analytical critique of private property relations and exploitation as an explanation of social inequality with a theory of moral and panhistorical "corruption." In

Hardt and Negri, "evil" is understood as the "corruption" of love and the common or what they elsewhere call "love gone bad":

> Our proposition [. . .] is to conceive of evil as a derivative and distortion of love and the common. Evil is the corruption of love that creates an obstacle to love, or to say the same thing with a different focus, evil is the corruption of the common that blocks its production and productivity. Evil thus has no originary or primary existence but stands only in a secondary position to love.[54]

According to Hardt and Negri's logic, since love is an ontological category that constitutes "being" and at the same time is a creative life force that brings into being new social forms, "love" brings about all social relations. Thus, social relations which are unequal are, so the story goes, best "explained" as the "corruption" of love or what they also call "love gone bad." To this end, they suggest that "As soon as we identify love with the production of the common, we need to recognize that, just like the common, love is deeply ambivalent and susceptible to corruption."[55] On these terms, Hardt and Negri posit that "capitalism" too, like all social forms is at root made possible by love, but a "love gone bad."[56] Capitalism, in their view, is not an historical relation based on private property relations and the exploitation of labor-power—the theft of surplus labor in production. Rather, they argue that capitalism and its alienating effects are best explained as a form of "love gone bad," by which they mean as a form of "corruption."

More specifically, according to Hardt and Negri, the "corruption" of love is manifested in "identitarian love" or "love of the same" rather than "love of difference." Race love, nation love, patriotism, romantic love, marriage-couple love are, for Hardt and Negri, examples of "love of the same."[57] As an "antidote" to love of the same, Hardt and Negri expand upon the concept of "love thy neighbor," and following Nietzsche they argue that higher than love of neighbor is "love of the farthest."[58] Here Hardt and Negri reduce social transformation to moral platitudes and lessons in multiculturalism and "loving difference." In doing so they erase the fact that exploitation under capitalism is the root cause of inequality and, as well, exploitation is entirely compatible with "love of the farthest" as transnational capitalism goes "all over the globe" and "must settle everywhere, nestle everywhere, establish connexions everywhere" not only to expand its markets but to secure sources of exploitable labor-power to stave off declines in profit.[59] What Hardt and Negri then propose as the "solution" to "love gone bad" or "love of the same" and toward the "common" is not social transformation but a "mass exodus" from institutions of the family, the nation, the corporation, and so on. As a "force to combat evil," Hardt and Negri contend, "love now takes the form of indignation, disobedience, and antagonism. Exodus is one means

[. . .] of combating the corrupt institutions of the common, subtracting from claims of identity, fleeing from subordination and servitude."[60]

Hardt and Negri's argument for a "mass exodus" from "old/bad" forms of love is, as I explicate further below, an updating of the culture of capitalism—of the methods used to help reproduce the social relations of production founded on exploitation—and not a break from capitalism. However, before further examining Hardt and Negri's theory of "love," "evil," "corruption," and "exodus" and their ideological role in transnational capitalism now, it is first necessary to understand the genealogy of Hardt and Negri's theory in classic idealism. This is important both because Hardt and Negri overtly claim to be producing a "materialist" theory and because their theory is taken by others as a "new" "true" materialism. Materialism, broadly, is the explanation of the origin of existence on the basis of exclusively material relations and laws of motion. When Hardt and Negri claim that love is "ontological" and constitutive of "Being," this does not mean that they are working with a materialist understanding of love. Despite their claims for a "materialist teleology"[61] and a "materialist perspective,"[62] at its core Hardt and Negri's theory of love and the "common" is not actually an historical updating of materialism for new material conditions of production that are in the process of formation, but an ideological updating of classic Christian idealist (spiritualist) ontology in which the "real" is represented as grounded in the ideal or spiritual. This can be seen not only in their direct references to St. Paul's theology, but in how closely Hardt and Negri's theory of "love as being" and "love as a force to combat evil" resembles Augustine of Hippos' classic Christian idealist ontology of "God as being" and of "good" and "evil" and their "relation" to each other. In classic Christian ontology, "God *is* being" and is the "real." More specifically, the concept of God, in classic Christianity, is understood as *absolute spirit*: the all-powerful, immutable, ineffable, excessive, and unquantifiable divine life force. This concept of absolute spirit, moreover, is regarded to be the absolute basis of the real as such and is regarded as "supremely good." In this theory of being, the concrete and sensuous world of the earth is considered to be an effect of God. The world, in other words, according to this ontology is "God's creation" and God is its "life force." However, according to classic Christian ontology, "God" is regarded to *compose* or *create* but *is not contained in* the world. In other words, for Augustine, the divine creator is *not* the creation. To put this another way, in classical Christian ontology, the world is not seen as "God incarnate" because, in this theory of ontology, "God" is an immaterial, divine force which cannot be incarnated. In this view, "God" as absolute spirit is outside any terms of empirical measurement, not contained in or by any material, concrete, or sensuous form or relation. In the *Confessions*, for example, Augustine remarks that "truth says to me: Your God is not heaven or earth or any kind of bodily thing."[63] Although, in the course of the *Con-*

fessions, Augustine formally rejects the theories of Plato and ancient Greek philosophy in general, his theory of "God" at the same time derives from Plato's concept of ideal forms and the theory of ontology of which it is a part. In *Republic*, Book X, for example, Plato distinguishes between an "ideal form," which he argues is the product of "God" and is the one "true" form; the concrete or material form manufactured by an artisan, which Plato regards to be a copy of the ideal form; and the image or representational form (as in painting or poetry) produced by what Plato regards to be an "imitator" of the concrete. In this theory of ontology the *ideal* form is the basis of being. The concrete form produced by human labor is considered to be a lesser copy of the ideal form produced by God, and the image or representation of the concrete form is regarded by Plato to be an even lesser copy of a copy—an "imitation" of a copy—of the ideal form.[64]

In the *Confessions*, "God" is being, and is the "real," and the concrete and sensuous world is a second order reality or "becoming" but not being itself. For example, Augustine remarks, "I considered all the other things that are of a lower order than yourself, and I saw that they are not absolute being in themselves, nor are they entirely without being. They are real in so far as they have their being from you, but unreal in the sense that they are not what you are."[65] In this view, "God creates the world in his own image" but the "image" (which in this theory of ontology is the concrete and sensuous world) is an *effect* and (lesser) *copy* of "God." The world then, because it is regarded to be created by a divine life force that is "supremely good," is also good, but because it is not this divine force in itself, it is a "lesser" good or second order good. As such, it is regarded to be subject to corruption. This "corruption" in Augustine is given the name of "evil." In the classic Christian ontology because God is regarded as absolute reality, as *Being*, and as supreme good, "evil" is understood not to "exist." To this end Augustine remarks:

> For you evil does not exist, and not only for you but for the whole of your creation as well, because there is nothing outside it which could invade it and break down the order which you have imposed on it. Yet in separate parts of your creation there are some things which we think of as evil because they are at variance with other things. But there are other things again with which they are in accord, and then they are good. In themselves, too, they are good.[66]

Evil, for Augustine, is the *absence of Being*—its disintegration—and the corruption of Being and the good. "Evil," in this narrative, is not an autonomous "force"—does not "exist"—but is manifested in the corruption, disorder, decay, and dissolution of God's creation which is good when things are "in their proper place" but evil when it becomes disordered.[67]

Hardt and Negri formally distance their theory of love (as divine life force) and evil (as earthly corruption of divine life force) from Augustine's theory of love, God and evil, by drawing from Spinoza's understanding that "evil" does exist but has a second order existence. They claim that,

> Spinoza's difference resides at a deeper level where the education or training of the mind and body are grounded in the movement of love. He does not conceive of evil, as does Augustine, for instance, as the privation of being; nor does he pose it as a lack of love. Evil instead is love gone bad, love corrupted in such a way that it obstructs the functioning of love. [. . .] And since love is ultimately the power of the creation of the common, evil is the dissolution of the common or, really, its corruption.[68]

This is not a break from the basic spiritualist ontology of Augustine. While Augustine claims that "evil" does *not* exist, he still deploys "evil" as an explanatory concept in place of a materialist explanation of historical relations of production. In like manner, when Hardt and Negri, following Spinoza, claim that "evil" *does* "exist" they are deploying a concept that vacates materialist critique.

The point of contention, however, is not the *specific* theory of "evil" but the use of the concept of "evil" to explain both material contradictions and the social inequality that arises from these contradictions. Any use of the concept of "evil" to explain capitalism and exploitation is an ideological mystification of material relations regardless of which specific theory of "evil" is used. The concept of "evil" ideologically translates the material contradiction of class society—and the historical and material causes of intensified economic exploitation and social alienation in private property relations—into an eternal existential condition of life as such. "Love," according to this ideology, is an autonomous life force but it is always already subject to "evil" and corruption. According to Hardt and Negri, "this is not to say we should imagine we can defeat evil once and for all—no, the corruptions of love and the common will continue."[69] The structural and systemic violence of capital, its onslaught into more levels and areas of human existence, and the social alienation that results from private property relations are abstracted from the historical conditions that produce them and transcoded into a transhistorical "corruption" rather than historical, and therefore, transformable material relations. As a result, as (neo-)classical Christian idealists, Hardt and Negri's theory also returns to Plato's "ideal forms." "Love" and "the common" are deployed as ideal forms in which training is needed while the struggle to transform the social relations of production—the material relations of capitalism—is considered a corrupt copy of an ideal form. This is the marketing of a new religion to the contemporary global working class, dressed up as material resolutions to the class contradictions workers face.

This does not *explain* but *explains away* and disappears the material contradictions in which workers live. Deploying the concept of "evil" to explain away material relations puts the production of materialist knowledge of objective relations in suspension—knowledge which is necessary for bringing about more effective collective social transformation—and instead is aimed at rallying people around "moral certitude" without knowledge. It is quite telling for instance, when Hardt and Negri argue that their Spinozan theory of "evil" provides a more effective "explanation for why at times people fight for their servitude as if it were their salvation, why the poor sometimes support dictators, the working classes vote for right-wing parties, and abused spouses and children protect their abusers."[70] "Such situations," they contend:

> are obviously the result of ignorance, fear, and superstition, but calling it false consciousness provides meager tools for transformation. *Providing the oppressed with the truth and instructing them in their interests does little to change things.* People fighting for their servitude is understood better as the result of love and community gone bad, failed, and distorted. [. . .] People are powerfully addicted to love gone bad and corrupt forms of the common. Often, sadly, these are the only instances of love and the common they know![71]

Hardt and Negri update a ruling class paternalism that workers do not need to know the truth of material relations or struggle to produce concepts that can most effectively explain these relations. According to their narrative, even if the oppressed "know the truth" of material relations and their objective interests, it does "little good" because the oppressed are "addicted" to their own oppression. Here, they displace the materialist concept of *exploitation* of the surplus-labor of the working class *by capital* as an explanation of material contradictions in capitalism, including poverty, with the psycho-affective concept of the "addiction" of the working class to "corrupt" forms of love. They displace the explanatory concept of "false consciousness" with the "ignorance, fear, and superstition" of workers. Hardt and Negri claim that their approach embraces the "power" of the poor. "The poor," Hardt and Negri contend, "are actually extraordinarily wealthy," by which they further elaborate: "despite the myriad mechanisms of hierarchy and subordination"[72] they are "creative" and "express an enormous power of life."[73] This is an ideological inversion of the exploitation of workers' surplus-labor under capitalism and a concealing of the brutal material contradictions under which workers live with the promise of "hope" and "spiritual wealth" in place of material equality. Moreover, according to this logic, workers in transnational capitalism do not suffer from the exploitation of their labor or poverty at all, they suffer from a pathology—from a state of "addiction"—and get in the way of their own already existing freedom and "extraordinary wealth." The oppressed and exploited, in other words, are represented as "pathological" and as having "chosen" their own poverty and exploitation. The implication of their argument regarding domestic violence, for instance, is that domestic violence is the "addiction" of abused spouses and children to "bad forms of love." This is not only not feminism, it is *sexist anti-feminism* which pathologizes the abused, oppressed, and exploited, and ideologically disappears the structural relations and the material contradictions of private property that enable the growth of domestic violence. What Hardt and Negri advance in the name of the "common" is a cynical voluntarism and moralism against the working class. Workers' collective struggle for socialism and to end the exploitation of their surplus-labor is demonized as "evil" and "corrupt" and what the workers "really need," according to Hardt and Negri's sermon, is education and training in "love." The (ideo)logic of this is that the exploited and oppressed are poor and exploited because they don't know how to love.

As in all forms of ideology, Hardt and Negri's biopolitics is enabled by specific historical and material relations and, at the same time, ideologically inverts them. While the theory of "evil" and "corruption" has a genealogy in classic Christianity, it is resurrected today as a managerial strategy for transnational capitalism. Hardt and Negri's theory of the "corruption" of love and the family and their argument for a "mass exodus" from the capitalist family is not so much a break from capitalism as an ideological updating of it. In order to critique Hardt and Negri's theory that the family is a "corrupt" form of love, it is next important to understand their argument that capitalism is a form of corruption. As I discussed earlier, in *Commonwealth,* Hardt and Negri advance their longstanding claim that "labor is increasingly autonomous" from capital and that "capital is increasingly external to the productive process and the generation of wealth."[74] "Biopolitical labor," Hardt and Negri claim, autonomously produces common wealth. "Capital" they argue, "may constrict biopolitical labor [and] expropriate its products [but] does not organize *productive cooperation.*"[75] In other words, according to this narrative, there is no *structural* relation of exploitation between capital and labor, therefore, the private appropriation of socially produced wealth by capital is *incidental* not systemic. It is in this way that Hardt and Negri go on to redefine capitalism not as a *mode of production,* but as a *form of "corruption"* of an already autonomously produced commonwealth. This is entirely consistent with mainstream corporate theories which explain away the global economic crises of capital as an effect of the corruption of a handful of rogue financiers and in so doing conceal the structural relations—the social relations of production in capitalism—which reveal that exploitation and crisis is systemic and endemic to production for profit, founded in private ownership of the means of production.

In much of their discussion of "commonwealth" what they are actually arguing for is a capitalist managerial strategy of raising the productivity of workers without transforming the social relations of production as a response to economic crisis. If capital is "external" to production and, moreover, production and labor are "autonomous" from capitalism then, according to this narrative, there is no need to transform the social relations of production and end private ownership of the means of production. Simply raising productivity itself—what Hardt and Negri call "expanding the commonwealth"—is, therefore, represented in *Commonwealth* as a "revolutionary" act of the "common." Their main concern is that:

> Labor-power has always exceeded its relation to capital in terms of its potential, in the sense that people have the capacity to do much more and produce much more than what they do at work. In the past, however, the productive process, especially the industrial process, has severely restricted the actualization of the potential that exceeds capital's bounds.[76]

By revising Marx's concept of "labor-power" with Spinoza's concept of *potenza* they naturalize the productivity of labor as an ahistorical "life force"—a natural and unlimited capacity of the body—and obscure the way in which individual concrete labor (work for wages) is shaped by abstract social labor (productive labor or what Marx calls "species-being"). In doing so, by arguing for the "actualization of the potential" of labor power to produce beyond the limits imposed by the current working day of capitalism and by arguing against the necessity to materially abolish class relations, Hardt and Negri produce an argument for the indefinite extension of the working day—seeking an increase in the rate of exploitation of the surplus-labor of the worker through an absolute extension of the working day. On the one hand, their argument (i.e., that labor power has the potential to "do much more and produce much more" than what is done at work) is concerned with increasing the extraction *by capital* of the surplus-labor of the worker by intensifying exploitation and extending the workday of the worker beyond the official working day in which they are paid an hourly wage. On the other, it articulates material interests of capital to constantly update its "production process" and techniques in order to remain competitive with other capitals and to eliminate outdated production practices that become a hindrance to the extraction of greater amounts of surplus-labor for profit. The actual aim of this corporate theory is not to end the exploitation of labor, but to get rid of the old *forms* of capitalist production and the old institutions that have supported its reproduction that are now becoming fetters to *profit for transnational capital*, without abolishing the *content* of capitalist production: private ownership of the means of production and the theft of workers' surplus-labor by the owners of the means of production (exploitation). This is another way of saying that *Commonwealth* supports increases in the rate of exploitation of workers in capitalism and represents intensified exploitation of workers as if it were liberation.

It is on these terms—i.e., on the terms of updating the relations of reproduction in capitalism and not on the terms of actually emancipating people from capitalism through ending the private property relations on which the family is founded—that Hardt and Negri declare that the family is a "corrupt" form of love and propose a "mass exodus" from the family. A telling example of the class politics of Hardt and Negri's concept of "new" love relations is their biopolitical fable of "wasp-orchid love."[77] In this fable, which they present as an mass exodus from capitalism itself, they draw from Deleuze and Guattari's discussion of "wasp-orchid machine" (*A Thousand Plateaus*) and rework it into a fable of "wasp-orchid love"—a concept of so called "new" love relations which they contrast with both "marriage, family, and couple love" and with "love of worker bees."[78] Hardt and Negri begin by criticizing Bernard Mandeville's eighteenth century entrepreneurial morality tale *The Fable of the Bees*, which is the story of a wealthy and prosperous bee hive "riddled with all order of private vices" and competitive individualism that, in the course of the fable, is then transformed into a "virtuous" and moral hive, only to fall apart economically.[79] More specifically, they criticize a pervasive interpretation of this fable, informed by Adam Smith, who reads it as a "confirmation of capitalist ideology."[80] Hardt and Negri particularly take issue with Adam Smith's understanding that "self-interest" is the basis of public good. "What Smith bans most adamantly from the marketplace," they argue, "is the common: only from private interest will public good result."[81]

In order to formally distance their own argument from this private property conception of human relations, they initially contrast Mandeville's pro-capitalist narrative, with the fable of the "love of worker bees and flowers."[82] This latter fable references the fact that bees, in the process of collecting nectar from fruit trees to produce honey, cross-pollinate fruit flowers, thus enabling them to bear fruit.[83] "The economic fable of these bees and flowers," they suggest, is not based on competitive individualism and "self-interest" but, by contrast, "suggests a society of mutual aid based on positive externalities and virtuous exchanges in which the bee provides for the needs of the flower and, in turn, the flower fulfills the bee's needs."[84] And yet, according to Hardt and Negri, while "Bees and flowers do indeed suggest a kind of love," it is, they contend, "a static, corrupt form."[85] This is because, they claim: "The dutiful worker bees [. . .] joined with their flowers in a virtuous union of mutual aid, are the stuff of socialist utopia. All of these bees, however, belong to the bygone era of the hegemony of industrial production."[86] Hardt and Negri's casual dismissal of "love of worker bees" is an oblique reference to the writings of Alexandra Kollontai in such texts as her novel, *Love of*

Worker Bees, as well as her theoretical writings and speeches such as "Sexual Relations and the Class Struggle" in which she develops a concept of "love-comradeship" or "red love" founded on the material abolition of class. Their rejection of production for need as "utopia" is already an indication of how much their narrative is steeped in capitalist triumphalism which presumes that class is the bedrock of the real. On this basis, for example, they not only rationalize capitalism as "love gone bad" but, they also demonize socialism and workers collective struggle to transform the material relations of production in capitalism "as evil"—as a "corruption of love and the common."[87]

Switching species in order to write a "new fable," Hardt and Negri turn to Deleuze and Guattari's discussion of the "wasp-orchid machine" in which, as Guattari puts it: "wasps fuck flowers."[88] More specifically, some orchids give off the odor of the sex pheromone of female wasps and moreover, their flowers are shaped like female wasp sex organs. In this case, pollination occurs not through the collection of nectar to produce honey, but through "pseudo-copulation" as "male wasps move from one orchid to the next, sinking their genital members into each flower and rubbing off pollen on their bodies in the process."[89] What makes this fable a "radical break" from the "family" according to Hardt and Negri, is its "immaterial production." Wasp-orchid love, they claim, "is a model of the production of subjectivity that animates the biopolitical economy."[90] Toward this end, they claim that: "These wasps aren't your dutiful worker bees; they aren't driven to produce anything. They just want to have fun."[91]; "The bees and flowers produce honey and fruit, but the wasps and orchids are just hedonists and aesthetes, merely creating pleasure and beauty!"[92] Moreover, they claim that "wasp-orchid" love evokes images of "serial sex" and "cruising," which they contend is a form of "resistance" to "marriage-couple and family love."[93]

What Hardt and Negri put forward is actually a private property conception of love. What Hardt and Negri call "love of worker bees," while an oblique reference to the writings of Kollontai, is actually an obscuring of Kollontai's Marxist theory of love and sexuality. What Hardt and Negri call "love of worker bees" in order to dismiss Marxist theory and socialism, is not the historical materialist theory of love and sexuality that is offered by Kollontai, but as evidenced in their story of "bees and flowers" is more or less a narrative of "bioharmony." By contrast, Kollontai's theory offers an enabling critique of the bourgeois family in its "nuclear" or "marriage-couple" forms that also has profound implications for a critique of the opportunist apologetics for capitalism that we see in the "post-family" writings of Hardt and Negri and their fable of "wasp-orchid love." Kollontai's critiques of "marriage-couple love" is not limited—as is Hardt and Negri's criticism—to a *moral* criticism of the family and what (as I discuss further below) is essentially a spiritual, cultural, and moral updating of ruling class "family" values into ruling class "(post)family" values. Rather, Kollontai offers a critique of the economics of the family that cuts through family values.

As I remarked earlier, for Kollontai, love is a "social emotion" the form of which develops as an effect of the material relations of production. Kollontai argues that love is an "integral part of culture"—meaning that it is part of the role of culture in reproducing the existing social relations of production. Love, in other words, is historical and does not transcend the social relations of production rather, love relations develop in relation to the social relations of production: "From the very early stages of its social being, humanity," living under specific historical relations of production, "has sought to regulate not only sexual relations but love itself."[94] This is because love serves to strengthen the bonds of relationships that become economically necessary to preserve the prevailing relations of production. This does not, of course, mean that the forms of love that prevail serve the interests of all of humanity rather, they serve the dominant class interests.

It is on this basis that Kollontai further critiques the traditional "marriage-couple love" of the bourgeois "nuclear" family form—with its proprietary culture of possessive individualism—that emerges under capitalism. Kollontai argues that this form of family has historically been useful for capital as a means of regulating the concentration of wealth into fewer hands and of privately reproducing the labor force out of the wages of workers, in the interests of profit for capital. Yet, while workers are economically compelled to live in the family and pool resources, the bourgeois family form does not actually provide freedom for workers. Just as private property relations give rise to a concentration of wealth within some families—and the family as an institution developed as a protection of private property relations—so it leads to poverty in working class families whose members are required to shoulder the burden of the cost of reproducing living labor. As Kollontai argues: "The destructive influence of capitalism destroys the basis of the worker's family and forces him unconsciously to 'adapt' to the existing conditions."[95] What she means by this is not that it destroys the *morality* of workers and "corrupts" them, but that it destroys the *economic* basis of their existence and impoverishes them. These are part of the irreconcilable class contradictions of capitalism: The family form that is useful for the ruling class to reproduce the concentration of capital into fewer hands becomes a space of dire need for the workers who must "privately" shoulder the cost of reproducing exploitable labor power out of their wages. Under the economic compulsion of class contradictions, workers find ways to adjust to the economic contradictions and develop new forms of love, kinship, and sexuality to work within the confines of capitalism. Kollontai, in her own historical moment, cites the development of love and sexual relations outside of the prevailing "marriage-couple" arrangements from the mere postponement of marriage and children, to adultery and prostitution, to polygamous marriage of three or four persons, to "civil" or common-law marriages, to "free love," and so on. Today these forms include not only "serial sex" and "cruising" (which are both cited by Hardt and Negri as an exodus from the family[96]) but various "post-nuclear" family, kinship, love, and sexuality forms such as divorce-extended families, transnational love relations, poly-sexual families, multi-ethnic families, post-marriage families, virtual and online relationships, urban tribes and, as I discuss further below, "transspecies" families among others. But these cultural shifts in the affective relations of the family, love, and sexuality do not in themselves constitute a transformation of the conditions that exploit workers to begin with or that give rise to the bourgeois family form. Without transformation of production relations, cultural shifts in love and sexuality relations are not freedom from necessity but the inevitable results of conditions of necessity—in short they are *adjustments to the contradictory needs of capital and the pressure that irreconcilable class contradictions exert on workers.* While Kollontai rigorously critiques the "hypocritical morality" of "marriage-couple love" under capitalism, the "root problem" with "marriage-couple love" is not simply its "hypocritical morality" but the "structure of [the] exploitative economy" upon which the family rests.[97] The "hypocritical morality" of the bourgeois family is not an effect of "moral corruption"; rather, it is a translation into culture of its underlying class contradictions in production. These same class contradictions undergird the so called "new" post-nuclear or post-family family. To bring about changes in the moral and cultural relations does not in itself change the exploitative economy upon which the family rests.

In fact, Hardt and Negri's theory conceals that transnational capitalism necessitates new and different forms of love, sexuality and reproduction relations. It is not workers who adjust out of "free consent"; rather, capital economically compels workers to do this in order to have a more easily exploitable workforce. The ruling class, Kollontai argues "seizes upon the new" family forms developed by workers.[98] Hardt and Negri's celebration of "wasp-orchid" love and their rejection of collectivity and solidarity of workers (under the rubric of "love of worker bees") is just as much a private property concept of love as "marriage-family couple love." Its reliance, for instance, on "cruising" and "serial sex" as "antidotes" to the bourgeois family form of "marriage-couple and family love" is not a break from private property relations or love and sexuality under the capitalist mode of production, but a marker both of the extension of the market further and further into human sexual and love relations—that is, the commodification of love and sexuality—and, moreover, of historical shifts in production practices, but not production *relations,* which put pressure on old family forms as an institution useful for reproducing capitalism. "Serial sex" and "cruising" are contemporary examples of what Kollontai calls the "passive adjustment of the working class to the unfavorable conditions of their existence."[99] Like "prostitution," serial sex and cruising with their market logics of "free and equal exchange" undergirded by structural relations of exploitation, are ruling class ideological resolutions to the material contradictions and social alienation faced by the working class under private property relations.

Kollontai's materialist analytics of the class contradictions of love under capitalism teaches workers not to be fooled by the ruse that the ruling class puts forward in equating changes in cultural mores and family values—which capital necessitates of workers—with changes in material relations. She points out that "The champions of bourgeois individualism" routinely "say we ought to destroy all the hypocritical restrictions of the obsolete code of sexual behavior."[100] They do so when they find that older codes of sexual behavior begin to become fetters for profit. And this is the case today with the "marriage-couple" arrangement which, in many instances, has come to serve as a hindrance to profit for capital. The key, for capital now, underlying the "new" post-nuclear forms of sex and love is that they are *flexible* families and kinship relations. That is, they are families that are flexible in their practices and schedules, in their gender and sexual relations, in their understanding of what "companionship" and "love" are—in short, they are flexible in the methods they use to reproduce labor-power to allow for greater pliability of the workforce to the dictates of production for profit in global capitalism. "Post-nuclear" forms of love, kinship, and family are commodified forms of love *not* in the sense of being morally "corrupt" as conservatives argue. The "new" forms of family are as commodified as the old forms because these family and love relations are shaped by the needs of capital today for a "flexible" labor force that has no permanent ties or commitments that will get in the way of higher and higher levels of exploitation. Capital needs living labor, for example, to be able to pick up and move for a new job or position in another city, state, or nation.

The decisive difference in Kollontai's historical materialist understanding of love and "love-comradeship" is that Kollontai's analysis does not conflate changes in forms of love with social transformation. "Love-comradeship" for Kollontai is an important social emotion in revolutionary struggle, but it is not an end in itself and does not "constitute" social transformation. The assumption that it does leads to the idea that changing *family values* automatically and spontaneously brings about material change and freedom from exploitation. Hardt and Negri put forward the understanding that "mass exodus" from the family is *in itself*—spontaneously and automatically—constitutive of transformation of material relations. Hardt and Negri's call for a "mass exodus" from the family, in which at the same time they actively deny the need for workers to work to transform production relations upon which social reproduction under capitalism rests, covers over the dire conditions of necessity under which workers are economically compelled to live in capitalism and works to dismantle struggles for social transformation, displacing transformation with cultural reform in capitalism. The "new" post-nuclear family under capitalism is a cultural updating of the "old" form family but not a material transformation of conditions of exploitation. It continues to be what Jen Roesch, following Engels, calls an "economic unit" under capitalism:

> The institution of the . . . family as an economic unit is central to meeting the needs of capitalism. Under the current system, employers pay workers a wage, but take no responsibility for most of the social costs of maintaining the current generation of workers—or for raising the next generation of workers into adulthood. Rather than these responsibilities being shared collectively by society as whole [. . .] they are shouldered by individual families.[101]

Under capitalism, workers reproduce their labor-power privately out of their wages while the majority of socially produced wealth is concentrated into fewer hands and put back into production for the profit of the owners. But a change in the gender, sexuality, marital status, and/or number of sexual partners and/or children is not a transformation of this material relation of capitalism in which the working class shoulders the economic burden of reproducing their own labor-power and the next generation of labor-power out of their wages while capital profits from the exploitation of their surplus-labor. Moreover, in all cases the worker still must give her wages back to the owners of the means of production in exchange for means of subsistence on top of submitting to exploitation for the wages to begin with.

What the narrative of "exodus" supports is not freedom of the worker from conditions of necessity—which would require transforming production relations and abolishing production for profit—but the unfettered onslaught *of capital* and the increased atomization of the worker. It is an articulation of the fact that capital, at its basis, does not need the "marriage-couple" form per se, so long as workers individually shoulder the cost of their own reproduction. What capital aims to do is to control the rate of growth and development of the surplus-value producing population. Thus when the "marriage-couple" is useful for reproduction of supply of workers in absolute terms it is promoted, but it can also get in the way of profit when capital needs an immediate supply of readily available labor to exploit—and thus capital works to dismantle or deregulate it and to produce reproduction, love, and sexual relations which are more and more subordinated to commodity exchange and production for profit.

Ultimately, Hardt and Negri's arguments are not "resistance" but a normalization of the contradictory needs of capital. On the one hand, they make an argument for the raising of productivity without the transformation of social relations of production. They endorse the notion that a "virtuous cycle" and raising productivity can "expand the common."

This is, as I have marked, a way to raise the rate of exploitation. It is a recognition—by capital—that the exploitation of labor is the basis of wealth and that no amount of financial/usury/immaterial production actually produces wealth and that to stave off economic crisis and the decline in the rate of profit ultimately requires increasing the rate of exploitation, one way of which is to increase the productivity of workers and get rid of old production practices that now stand in the way of profit for capital.

On the other hand, Hardt and Negri also maintain that the exploitation of labor is part of a bygone era; that material production has been superseded by immaterial—or biopolitical—production and that what is important about biopolitical production is not the production of material needs but the production of subjectivities. In their reading of "love" and endorsement of the fable of "wasp-orchid love" they proceed to oppose the notion of a "virtuous cycle" which they now situate as part of the "love of worker bees." What they are "discontented" with regarding the latter is its focus on "material production." What they like about "wasp-orchid" love, by contrast, is that it focuses on "immaterial production" of non-productive affect, beauty, pleasures, fun, and "subjectivity"—the sensual values of the market.

This contradiction—on the one hand raising material productivity, without ending exploitation, in the name of "commonwealth" and on the other hand rejecting "productive labor" in the name of immaterial production—is the effect of the contradictory needs of capital. On the one hand, capital needs to raise the material productivity of workers in order to extract more surplus-labor precisely because the "immaterial economy" (concentration of service and non-productive labor in the North as well as the export of capital to the South) has not led to increases in wealth rather it is related to the crisis of profit. On the other hand, capital works to conceal the fact that it is based on the exploitation of productive labor and does so by ideologically translating material contradictions—which are the result of the exploitation of labor—into cultural values. The worker must produce more and more material wealth but expect affect in return—she must be contented with increased exploitation and poverty as long as she is "rich" in "love" or "sex," etc.

Capitalism's Other-Worldly Omni-Species

While Hardt and Negri articulate a general theory of "biopolitics," the development of biopolitical ideology in contemporary cultural theory takes on many forms. One such significant form is the development of theories of transspecies posthumanism, which is articulated in the writings of Derrida (*The Animal That Therefore I Am*), Giorgio Agamben (*The Open: Man and Animal*), Donna Haraway (*When Species Meet*), among others. Like other forms of biopolitical ideology, transspecies posthumanism is also characterized by the displacement of "class," "labor," and production, with "life" and "life force" (e.g., the "power of life" etc.) as "explanations" of material relations of production. More specifically, however, transspecies posthumanism goes through the relay of the "animal" and "multispecies epigenetics" in order to ideologically displace "human labor power," the dialectical praxis of labor and the social relations of production. Contemporary posthumanism, to put this another way, is a theory of *transspecies*—an obscuring of the evolution of humans (through their labor) from non-humans—e.g., "animals." Humans are, of course, animals but they are animals with an historical difference—they have developed a form of reasoning which is itself developed by praxis, that is, by the dialectical praxis of labor and the ensemble of social relations of production within which it develops. One of the main goals of contemporary posthumanism is to divert attention away from class relations and exploitation of surplus-labor, by enacting a "fissure" in the concept of the human—that is, by ideologically dissolving the historical difference between human and animal—and, in so doing, invoking a "crisis" in the concept of human-labor power.

Central to this movement in cultural theory is Donna Haraway's book, *When Species Meet*, in which she argues for the dismantling of what she calls "The Great Divides" between animal and human. Haraway cites as evidence of the collapse of boundaries between human and other animals, "the fact that human genomes can be found in only about 10 percent of all the cells that occupy the mundane space I call my body; the other 90 percent of the cells are filled with the genomes of bacteria, fungi, protists, and such, some of which play in a symphony necessary to my being alive at all, and some of which are hitching a ride and doing the rest of me, of us, no harm."[102] It is, for example, on this basis that Haraway puts forward the understanding that there is an *aporia* in Marx's concept of human labor power. She contends that while "Marx understood relational sensuousness, and [. . .] the metabolism between human beings and the rest of the world enacted in living labor [. . .] he was finally unable to escape from the humanist teleology of that labor—the making of man himself."[103] "In the end," she argues, "no companion species, reciprocal induction, or multispecies epigenetics are in his stories."[104] In short, Haraway's argument is that Marx has left out the "animal"—more specifically, the genetic constitution of the "human" by various "companion species"—and, therefore, according to Haraway, his focus on human-labor power is violently "anthropocentric." According to this transspeciesist logic, the biological constitution of the body defies the social logic of capital because on the one hand, the "animal" cannot be understood in terms of exploitation, Haraway claims, because "paws" are not "hands." On the other hand, human bodies are "constituted" by myriad microscopic "companion species" and therefore cannot be said to be distinctly "human" at all.[105] The exploitation of the surplus labor of the majority of workers by a minority of owners in capitalism is, by this logic, evidently "biologically impossible."

This is a rehearsal and updating of the ideologic of eighteenth and nineteenth century pseudo-scientific discourses of biological determinism such as craniometry in which the *social and historical* relations of production that give rise to exploitation were obscured and ideologically naturalized through the relay of the physical "raced" body. To be clear, transspecies posthumanism is a radically different *form* of this ideology. Haraway, for instance re-maps the body not in terms of the "cranium" and "race" but in terms of "DNA" and "multispecies epigenetics." Moreover, while craniometry worked to ideologically construct all kinds of so-called sub-species, posthumanism, by contrast, ideologically dissolves historical boundaries between species. For transspecies posthumanism, it is not that human and other animals are "identical" but that *all species—and the material relations of evolution as well as the historical differences in the material conditions of evolution of a species*—are now ideologically dissolved into an "omni-species." This omni-species, so the story goes, is itself constituted by multiple and, above all, post-binary "differences" whose multispecies epigenetics manifest themselves in what are believed to be "undecidable" ways. Any conceptualization of historical boundaries between "species"—which would situate them in relation to the structural relations in which they are produced—are regarded to be an arbitrary and unethical cultural construction and a violent fixing of amorphous, undecidable, plural, micro-differences into metaphysical binary difference.

This is a change in the *form* of eighteenth and nineteenth century pseudo-scientific ideology but not its *content* which reverts to the "physical body" and the so called "received biology" of bodies in order to obscure the social relations of production. Haraway substitutes the *effects* of evolution for the dialectical relations of evolution when she displaces the *material conditions and relations* under which natural selection and the evolution of diverse species are made possible, with the genetic "constitution" of bodies. This is a matterist understanding of biology which substitutes the bodily sensuous—the received biology of "flesh and blood" or the "physical"—for the ensemble of material relations that determine the development of a species. But biological organisms are mediated by what Marx explains as the dialectical praxis of labor which acts on nature and transforms it in order to meet needs and in doing so produces new needs.[106] Even Haraway's favored animal—the dog—is not outside of the dialectical praxis of labor and the social relations of production in which it develops—and the way in which these structural relations transform the natural environment. As a species, the evolution of the dog from the wolf is itself the effect of the dialectical praxis of human labor as it intervenes in nature and transforms it to meet needs, and in doing so transforms its own needs. Dogs are a result of human breeding of wolves: "Humans lived as roaming hunters and gatherers for most of their existence [and] wolves began following hunter-gatherer bands to feed on the wounded prey, carcasses or other refuse. At some stage a group of wolves, who happened to be smaller and less threatening than most, developed a dependency on human groups, and may in return have provided a warning system. [. . .] Several thousand years later [. . .] people began intervening in the breeding patterns of their camp followers, turning them into the first proto-dogs."[107]

More generally, and more importantly, as Richard Lewontin and Richard Levins argue in *Biology Under the Influence*, "The socially conditioned [production] and transformation of our environments" through the dialectical praxis of labor and the social relations of production in which this labor is organized "determine the actual realization of our biological limits."[108] Biology, they argue, is not just "received biology" but "socialized biology" by which they mean it is mediated by the material relations of production. This is why, in advanced capitalism,

> A severe winter in an urban environment does not produce frostbite but hunger—when the poor divert resources from food to fuel. Racism becomes an environmental factor affecting adrenals and other organs in ways that tigers or venomous snakes did in earlier historical epochs. The conditions under which labor power is sold in a capitalist labor market act on the individual's glucose cycle as the pattern of exertion and rest depends more on the employer's economic decisions than on the worker's self-perception of metabolic flux [. . .] It is the social mediation of individual biological phenomena that turns a single day's incapacity from the flu into the loss of a job from an already marginalized worker, with consequent catastrophic economic failure and a [further] disintegration of health and general conditions of life.[109]

While Haraway proposes that transspecies posthumanism is a "materialist" theory which is more concerned with "mud," "slime," and "earth," than it is with the "sky" or "heavens,"[110] having ideologically displaced the historical conditions and material relations that enable material development of "received" biology, transspecies posthumanism—like other biopolitical discourses—puts in their place *spiritual* causes. By dissolving all species into an "omnispecies," transspecies posthumanism translates the material relations of evolution and history through which biology is socially mediated by the dialectical praxis of labor into a *generalized trans-history*. The theory of transspecies, to put this another way, ideologically dissolves *historical life*—life in historical and material relations of production—into general abstract "natural life" outside history. However, when the concept of "natural life" is abstracted from the historical and material conditions and relations in which it is produced, it is at root a spiritualist theory of life. This fundamentally idealist concept of life is part of the contemporary revival (also seen in the work of Agamben) of the concept that in classical Greek philosophy is understood as *zoē*. As marked further above, ancient Greek philosophy makes a distinction between two concepts of life: *zoē* and *bios*. While *bios* is regarded as mortal, biological life of, for example, the individual, *zoē*, by contrast, was understood as abstract general life (what is "common" to humans, animals, and the gods according the ancient Greeks) and more specifically as *transcendent*, spiritual "life" in the absolute. What makes possible living beings, by this logic, is not definite historical and material relations but an ahistorical, common "life-force," which is a concept of life that has its root in a spiritualist ontology (idealism).

If we turn to some of Haraway's cultural analysis, we begin to see this spiritualism (and particularly "animism") at work. For instance, her reading of "Jim's found dog"—a photograph, taken by a colleague Jim Clifford, of "a redwood stump covered with redwood needles, mosses, ferns, lichens. . ." that resembles the shape of "an attentive sitting dog"[111] —is premised on a spiritualist theory of "materiality" that is almost identical to the theory at work in the "found" images of the Virgin Mary in shadows, water stains, bird droppings—and in 2004 on a ten-year-old grilled cheese sandwich auctioned on eBay for $28,000. According to Haraway's narrative it is the evolved consciousness that recognizes the "canine soul [that] animate[s] the burned out redwood."[112] "Whom and what do we touch when we touch this dog?" Haraway asks, and she continues, "How does this touch make us more worldly, in alliance with all the beings who work and play for an alter-globalization that can endure more than one season?"[113]

Here, Haraway reads "Jim's found dog" as a figure of an "other" globalization, what she also calls "alter-globalization" or "autre-mondialization" and which she claims is a "more just and peaceful other-globalization."[114] But her theory of "other-globalization" actually has very little theoretical connection to the "worldly," or any materialist understanding of the world. Rather it is a signifier for a quasi-religious "other-worldliness." The "found dog" in Haraway's discourse is a trope for an ineffable, unrepresentable, transhistorical life-force. To touch the "found dog" is, in this narrative, to touch a life-force that "animates" all beings great and small from bacteria, fungi, protists, to the H1N1 virus containing swine, human, and bird genes, to the genetically altered onco-mouse and the cyborg. By this logic, an "other globalization"—that is, a new social form, an *other* world or other global society—does not have to be brought about by material transformation of historical relations in transnational capitalism. Rather, an "other-globalization" *already* exists *within* the existing. We merely have to come to recognition of our "natural co-existence," and it is this recognition that will actualize the "other-world" in practice.

The way to this "other-world," according to Haraway's narrative, is through transspecies "love." "Love" of multi-species is conceptualized in Haraway's argument as a transformative life force that helps us "be in touch" and "bring into being" the "other-world" within the world. Haraway contends that "love" is a "world making" activity that brings into being "other-globalization." It is "love" of "companion species," "messmates," etc. . . that, she argues, brings into being "a more just and peaceful other-globalization."[115] This is to say that, like Hardt and Negri, Haraway not only displaces "class" and the dialectical praxis of labor with "life" and "life force" but also regards love—in this case love of multispecies "others" which is a transspecies trope for what Hardt and Negri call love of difference and alterity or "singularities in the multitude"—as a creative force and "world-making" activity that brings into being new social forms. And yet, the "new social forms" that Haraway argues for are not a break from capitalist globalization; they are founded on it. For example, Haraway puts forward the idea that the emergence of the "transspecies family" and intimacy with pets under capitalism's commodity relations marks a "new" and more "evolved" set of family relations. More specifically, she puts forward a narrative in which the market and the commodification of pets and "pet needs" has developed a new and "improved" emergent transspecies family relation. Owing to the onslaught of the market and capitalist production into more and more levels of social existence and the profits to be made by capital off of the pet industry (from pet food, to chiropractic adjustments, mental health therapy, and prescription anti-depressants for pets), now animals—specifically the *pets of the privileged*—have "rights." Haraway remarks: "dogs in capitalist technoculture have acquired the "right to health," and the economic (as well as legal) implications are legion."[116]

In Haraway's narrative, capitalism has evolved and North Atlantic capital is the regime of the evolved who recognize their "kinship" with animals and call for new transspecies family and inter-subjective relations. For example, when comparing the cost of cholesterol medication for humans with the cost of "doggie dinners," Haraway moralizes that she would "throw away my Lipitor before I shorted my dogs and cats."[117] On such practices she contends, "No one can convince me that this [. . .] reflects bourgeois decadence at the expense of my other obligations."[118] "Furthermore," she adds,

> There could be no end to the search for ways to relieve the psycho-physiological suffering of dogs and, more, to help them achieve their full canine potential. [. . .] I am convinced that it is actually the ethical obligation of the human who lives with a companion animal in affluent, so-called first-world circumstances. I can no longer make myself feel surprised that a dog might need prozac and should get it—or its improved, still-on-patent offshoots.[119]

This narrative puts in suspension materialist critique of class relations and the social relations of production in which the needs of all sentient beings—human or animal—are subordinated to production for profit and in which working class families are economically forced to negotiate between basic needs for its members (whether mono or multi-species) such as food and medicine. It then translates these material contradictions of capitalism into "family values" and "ethical" consumer choices. To critique the class privileges of the North Atlantic consummative subject and the sentimentalizing of these privileges—which are effects of class contradictions originating in exploitation in transnational capitalism—as a manifestation of "bourgeois decadence" is, consequently, represented as anthropocentric and "unethical" to the treatment of animals. Haraway's argument is not at all a point of departure from "anthropocentrism" but rests on the very presupposition of human and *class* superiority—the evolved superiority of the first world subject—that it purportedly breaks from. Haraway's reading presupposes that it is in their ethical superiority over others, that *some humans* have the evolved consciousness to "recognize" their "kinship" with animals. This is a class narrative which ideologically sutures an "evolved consciousness" to the class privileges of some which are founded on material relations of exploitation and production for profit in transnational capitalism. This narrative uses animals as a *decoy* to disappear class relations and present social inequality as an effect of cultural and family values.

Haraway claims her argument is a "materialist" understanding of love, because she understands "love" as a "worldly" activity: "To be in love," Haraway remarks, "means to be worldly, to be in connection with significant otherness and signifying others, on many scales, in layers of locals and globals, in ramifying webs. I want to know how to live with the histories I am coming to know. Once one has been in touch, obligations and possibilities for response change."[120] The "becoming worldly" in Haraway's discourse is, like Hardt and Negri's "biopolitics," rooted in a spiritualist ontology in which abstract, transhistorical "life" is said to have "practical" consequences and bring about new social forms and new material relations. However, insofar as Hardt and Negri's biopolitical discourses draw from classic Christian ontology, the transspecies posthumanism advanced by Donna Haraway and others, by contrast, is perhaps closer to the idealist ontology of Hinduism manifested in such texts as *The Bhagavad-Gita,* in which the divine is said not just to *create* but to *constitute* the cosmos. Whereas in classic Christianity the divine creates but is not itself contained in the material and sensuous world or the cosmos, in ancient Hinduism the creator *is* the creation. In *The Bhagavad-Gita,* for example, Krishna says to Arjuna "I exist in all creatures"[121] and "all creatures exist in me."[122] In the *Gita,* "all creatures"—in fact, all aspects of the sensuous and material world, even inanimate and human-produced objects—are considered to be "fragments of divine power."[123] In this spiritualist ontology, the divine is thought to divide itself—*like a cell*—into different entities creating the appearance or "illusion" (*Maya*) of separateness when all, according to this understanding, is actually a *fragment* of the (same) divine absolute, the same spiritual force.

In the *Gita,* historical differences between "creatures" are dehistoricized and ideologically dissolved into a spiritual life force. Historical and social life—that is, the life of the individual or the species enabled by the ensemble of historical and material relations—is represented as an epiphenomenon of an eternal transcendent absolute spirit or spiritual "life force." In the *Gita,* for example, Krishna says to Arjuna, "Just as the embodied self / enters childhood, youth and old age, / so does it enter another body."[124] The basis of material life according to the *Gita* is not the historical or social life of the individual or even the historical life of the species, it is a transhistorical and transmaterial absolute spirit that goes through the eternal "cycle of many births."[125] The "embodied self" in this translation of the *Gita* is a reference to the Sanskrit concept of *atman,* what is roughly translated into the "true," transcendent and abiding (and for lack of a better term) "self." Unlike the Christian concept of the soul, or later Western concepts of the "self," however, *atman* is not considered to be a distinct individual soul or self. The *seemingly* individual soul (*atman*), in this understanding, is indivisible from infinite spirit or the "absolute" (*Brahman*). This is another way of saying that in this narrative it is a transhistorical or divine "absolute" that is the "life force" of the material and concrete world. Krishna continues: "Contacts with matter make us feel / heat and cold, pleasure and pain. / Arjuna, you must learn to endure / fleeting things—they come and go!"[126] Material reality is seen as an epiphenomenon of the divine absolute: "Nothing of non-being comes to be, / nor does being cease to exist; / the boundary between these two is seen by men who see reality."[127]

In the *Gita,* in other words, social history is under-written by a spiritual trans-history and, as well, there is an *"other-world"*—a spiritual and divine *Being*—that *exists within* and at the same time enables the material and concrete world (i.e., the creator *is* the creation). Moreover, the material contradictions, differences, and the social inequalities that exist are understood as epiphenomena of a spiritual and other-worldly common infinity. All are already part of this "other-world," the divine absolute is already in existence; however, according to the *Gita* people suffer from "delusion" and inaction and do not recognize their unity as fragments of the divine. They become attached to the "illusion" (*Maya*) of Brahman—to historical and concrete differences—and are "unaware" of their place in the other-world or spiritual absolute.

It is through this idealist ontology that *The Bhagavad-Gita* provides a spiritual resolution of material contradictions and, in doing so, it ideologically legitimates social and economic inequality, particularly caste. *The Bhagavad-Gita* ideologically converts the caste system and its economic and social inequalities which are the effect of an historically produced and therefore transformable social division of labor and production relations founded on exploitation into sacred "difference" and affirms different "sacred duties" (*dharma*) on earth in the name of "awareness" and "recognition" of one's specific "place" in the cosmos (i.e., as a "fragment" of a divine absolute). "Awareness," or the evolved spiritual consciousness as it is defined in *The Bhagavad-Gita* means, "action" (*karma*) with "detachment from the fruits of action."[128] More specifically, this means performing one's caste duty without questioning or focusing on the material conditions or material consequences of action. This is, after all, the basis upon which Krishna councils Arjuna to go to war with his cousins, when Arjuna questions the historical and material consequences of doing so, including the validity of killing his kinsmen for a kingship. "Action with detachment to the fruits of one's action" is often read as a criticism of self-serving individualism—*The Bhagavad-Gita,* for instance, emphasizes "detachment" from the "concrete" self and, as well, from material gain or the "fruits" of action. However, it is also a rejection of social critique and collective material transformation and an ideological legitimation of exploitation and the caste system. It is important to note that *The Bhagavad-Gita* itself is an historical document which is forged out of the history of class relations. As the sixth book of the *Mahabharata*—or "the great tale of the Bharata Dynasty"—its tales are part of the cultural reproduction, in the form of folklore and mythology, of the private property relations of ancient India. What Arjuna depicts as a brutal class war in which cousins are pitted against each other to slaughter each other and protect the private property holdings of competing dynasties, is represented in the *Gita* as the "sacred field of battle." In the context of private property relations and a caste-system which rests on private property, the command to perform one's duty but to remain detached from the "fruits" of action is an ideological legitimation of exploitation.

Moreover, *The Bhagavad-Gita* provides further ideological resolution to material contradictions through the concept of the "equal eye." In the *Gita,* the "aware person" and the person of discipline see past duality to the common spiritual life force of all: "Learned men see with an equal eye / a scholarly and dignified priest, a cow, an elephant, a dog, and even an outcaste scavenger."[129] While the *Gita* puts in suspension the possibility of material transformation, in place of material equality it puts forward a spiritual equality, the basis of which is "recognition" and seeing the "common" spiritual life that unites all creatures: "Arming himself with discipline, / seeing everything with an equal eye, / he sees the self in all creatures / and all creatures in the self."[130] It is in this way that the material relations of caste—for example, the division between mental and manual labor—and, as well, the class relations of the time (private ownership of the means of production) are ideologically preserved in the name of spiritual recognition and equality.

Transspecies posthumanism advanced by Haraway and others is an ideological updating of the ruling class ideology in *The Bhagavad-Gita.*

The ideological dissolution of historical differences between species and the positing of an "omni-species" or a "transspecies" in posthumanism is an updating of the *Gita*'s concept that "all creatures" are "fragments" of the divine absolute. Moreover, like the *Gita*'s concept of the "equal eye" which translates material inequality into spiritual equality, Haraway's argument for "transspecies love" is an argument for the spiritual recognition of a "common" life-force constituted by multiple differences, without social transformation of the historical and material relations of exploitation that bring about inequality. Just as the *Gita* is actually a form of ruling class ideology, so is transspecies posthumanism. What is most telling about the class politics of transspecies posthumanism is the way it sentimentalizes and naturalizes the social and economic inequality that arises from *historical and social* (and, therefore, *transformable*) relations of production founded on the theft of surplus labor (exploitation). At the core of Haraway's concept of "transspecies love" is what she calls "non-mimetic sharing," particularly "non-mimetic sharing" of suffering and pain: "Human beings' learning to share other animals' pain nonmimetically," Haraway remarks, "is in my view, an ethical obligation, a practical problem, and an ontological opening. Sharing pain promises disclosure, promises becoming."[131]

This concept naturalizes the class contradictions of capitalism as a transhistorical and inevitable part of *life as such*. More specifically, Haraway's theory of "non-mimetic sharing of pain and suffering" is based on the argument that in nature—in ecology for example—there are unequal or what she calls "non-mimetic" relations of use including the killing of some by others and, owing to this fact, there are no social relations that exist outside of "use" and "killing."[132] The *structural* relations of production in which use-values are produced through a process of exploitation, and killing occurs for profit, are ideologically translated into a generalized trans-history of the eternal and perpetual cycle of life, use, killing, and death. This is not unlike the grounds upon which Krishna chastises Arjuna for questioning why he should accept his position in the warrior caste by going to war and killing his cousins. On the terms of the *Gita*, while an individual soul (*atman*) might achieve transcendence (*moksha*) and permanent dissolution into the divine absolute without being reborn, there is no end to the cycle of samsara and even the divine cosmos itself has its "day" and its "night"—its own cycle of samsara. However, in contrast with the *Gita*, Haraway's argument is that humans must more fully and physically *feel* the pain of this process rather than detach from it, by involving themselves in the pain that is endured by other animals. This, however, bypasses social relations and restores what Roland Barthes criticized as the ideology of the "eternal lyricism" of life which de-historicizes birth, death, rebirth, pain, suffering, joy, and so on.[133] Haraway marginalizes the question of transforming the mode of production. Her argument distracts away from the necessity of collective social transformation of historical and, therefore, transformable, class relations and represents these structural relations as if they were permanent features of life. By her logic, since "using" and "killing" exists in all modes of production because it exists in nature, and the question of private ownership of the means of production is irrelevant, all we can do is the more "mundane" task of using ethically, killing ethically within capitalism. It is on these terms that Haraway argues: "Human beings must learn to kill responsibly. And to be killed responsibly [. . .]"[134] "Non-mimetic sharing" is an argument for so called "ethical" exploitation and so called "caring capitalism" whose actual aim is to naturalize and legitimate the expansion of transnational capitalism into all levels of existence. It is an argument for "equality" in *exploitation* for the majority.

As a figure for what she calls "non-mimetic" sharing of suffering, Haraway discusses the fictional character Baba Joseph, who is a lab worker in a scientific outpost in Zimbabwe in Nancy Farmer's novel *A Girl Named Disaster*.[135] As Haraway describes, Baba Joseph works with guinea pigs who are being used as part of sleeping sickness research to save the lives of humans and other animals. As part of this research the guinea pigs are kept in wire cages, shaved and painted with poisons and then exposed to tsetse flies who bite them and suck their blood. In the course of the research Baba Joseph puts his unprotected arm in the cages and, like the guinea pigs, his skin is extensively bitten and swells up. His reasoning for this: "It is wicked to cause pain, but if I share it, God may forgive me."[136] What Haraway finds so significant—and ultimately affirms—about the character of Baba Joseph is that he does not resign from his job and lose his status in the community, he does not try to convince a young girl, Nhamo, not to work in the lab, and he does not consider freeing the guinea pigs or tsetse flies.[137] In other words, Haraway celebrates the fact that Baba Joseph does not raise questions about the conditions in which he is working, or the structure which gives rise to these conditions, but subjects himself to physical pain and asks for forgiveness. Even more significant is the unsaid of Haraway's argument and its postcolonial reason: Baba Joseph also, evidently, does not question the relations that lead to his own exploitation in transnational capitalism. He is, for all intents and purposes, a "good" worker of the global South from the standpoint of transnational capital because he does his "duty," accepts the conditions of his exploitation in transnational capitalism, helps to skill the next generation of workers in the same, and if that were not enough he sees himself as the aggressor who must ask for forgiveness.

In this argument, all class contradictions are translated from structural relations into matters of personal responsibility, consent, and the "correctly" evolved moral and ethical values. Toward this end Haraway argues:

> Sometimes, perhaps, "taking the place of the victim" is a kind of action ethically required, but I do not think that is sharing, and further those who suffer, including animals, are not necessarily victims. What happens if we do not regard or treat lab animals as victims, or as other to the human, or relate to their suffering and deaths as sacrifice? What happens if experimental animals are not mechanical substitutes but significantly unfree partners, whose differences and similarities to human beings, to one another, and to other organisms are crucial to the work of the lab and, indeed, are partly constructed by the work of the lab?[138]

Haraway's concept of "significantly unfree partners"—which is a form of her argument for non-mimetic sharing—obscures the historical relations of production and the social division of labor. On one level, as marked above, this concept puts forward the idea that all are "significantly unfree" to transform social relations and are bound to an "eternal lyricism" of life (of birth, death, pain, love, joy). But, more significantly, Haraway deploys the concept of "significantly unfree partners" to represent lab animals and workers as having a "significant *partnership*" in the "unfreedom" of their labor relations. Her argument that animals and lab-workers are not "victims" but "partners" is by no means a "new" argument. Far from moving beyond "anthropocentrism," this is a form of social contract theory which argues that, at root, social relations are founded on agreement and mutual consent. Haraway projects onto *both* animals and exploited workers in labs *consent* to a global social division of labor in capitalism that is founded on production for profit and the exploitation of labor. Haraway contends that there are already "degrees of freedom" *within* the existing social relations of production. "Lab animals," she contends, "have many degrees of freedom [. . .] including the inability of experiments to work if animals or other organisms do not co-operate."[139] It is on this basis that Haraway reads the height of *alienation* under private property relations—what in colloquial terms is referred to as "losing the will to live"—as a "degree of freedom" when she argues that: "Even factory meat industries have to face the disaster of chickens' or pigs' refusal to live when their co-operation is utterly disregarded in an excess of human engineering arrogance."[140]

Haraway substitutes moral outrage for explanatory critique and social transformation and in doing so she uses the sensationalizing and moralizing rhetoric of "factory meat industries" as an "excess of human engineering arrogance" to detract attention away from the fact that her theory is a most effective ally of transnational capitalism. At every turn, Haraway obstructs any serious examination and critique of the structural relations of production—of production for profit—and the way they determine and structure existing "relations of use" between owners and workers, between the global North and the global South, between humans and other animals, between humans and natural resources:

> I resist the tendency to condemn all relations of instrumentality between animals and people as necessarily involving objectification and oppression of a kind similar to the objectifications and oppressions of sexism, colonialism, and racism. I think in view of the terrible similarities too much sway has been given to critique and not enough to seeing what else is going on in instrumental human-animal world makings and what else is needed.[141]

By suspending social critique for sentimentality, Haraway, echoing the *Gita*'s injunction against social critique, ideologically converts the social division of labor founded on class relations and production for the profit of a handful of owners into "cosmic" duties. In Haraway's narrative, the clinic-worker who accompanies animals to be euthanized, the "good" lab worker who has his hand eaten by tsetse flies without questioning the conditions of his own exploitation, the North Atlantic dog owner who puts her depressed pet on Prozac, etc., are all "fulfilling their ethical duty" to the cosmos and performing a "daily service of love," while persons who critique systemic relations of production and who argue for collective social transformation are described as anthropocentric, speciesist, reactive, lacking "response-ability."[142] Haraway formally distances her theory of "cosmic responsibility"—what she calls *cosmopolitics*—from the ideology of the *Gita* and other similar theories of divinity. She claims that her concept of the cosmos "is the opposite of a place of transcendent peace."[143] Haraway ostensibly rejects "transcendent unity" for what she refers to as a more mundane, worldly, and ordinary multiplicity and plurality. "The cosmos," she claims, "is the possible unknown constructed by multiple, diverse entities."[144] However, Haraway's ostensible rejection of "unity" and "transcendence" is not so much a break with the ideology of the *Gita* as it is an updating of its class politics for the class relations of transnational capitalism, which, in contrast to the private property relations manifested in the ancient caste system, require flexibility of contemporary workforces to be pulled in and out of different sectors of the workforce depending on what is profitable to capital. Given the *Gita*'s argument that a sacred duty (by which is actually meant caste duty) is one's place *in the cosmos* (i.e., and the cosmos *is* the divine whole as far as the *Gita* is concerned), Haraway's concept of open "cosmic duty" is only a thin ideological updating of the concept of a fixed "sacred duty" and its violent class politics—updated for an era of "green capitalism" and the "greening" or "bio-diversifying" of exploitation.

Non-mimetic sharing, to put this another way, is a euphemism for *exploitation*—for the theft of the surplus-labor of workers by a handful of owners. Moreover, as a euphemism, it is deployed to conceal and legitimate class relations. It is part of the ideology of imperialism. "Non-mimetic sharing" is an ideology that erases the relationship of difference and inequality around the world to the social relations of production founded on private ownership of the means of production and the theft of surplus-labor of workers around the world by a handful of owners. It affirms class inequality as "cultural difference" and erases that these differences are actually *in relation* to social structures of exploitation. All differences are relativized into a post-difference hegemony of "transspecies classlessness." According to this "transspecies classlessness," it is the "ethical" obligation—the cosmic, sacred, or "ethically evolved" duty—of the North Atlantic subject to give her dog prozac for depression and gourmet dinners, and to participate in the sport of fun agility competitions in her leisure time and, by contrast, it is the cosmic duty of the global worker of the South to have his unprotected flesh eaten by tsetse flies to pay for the "sins" of science and mankind. This is a legitimation of inequality which stems from private property relations and the exploitation of humans by humans. It is the *sine qua non* of what Franz Fanon so aptly critiqued in his title (and book) *The Wretched of the Earth*, in which differences that are the result of the social division of labor, the structural relations of exploitation in capitalism—and the violent transfer of wealth from the global workforce to a handful of owners around the world—are represented as "cosmic" duties or spiritual "callings" and, relative to the North Atlantic subject, it is the worker of the global South whose "cosmic duty" it is to do the most "non-mimetic sharing" of suffering.

Transspecies posthumanism has become a way not to free all species of animals from the excesses of the material contradictions of capitalism and production for profit but the latest ruling class strategy to dismantle struggles for social emancipation on the grounds that they are violently "anthropocentric." Socialism, feminism, anti-racism, the struggle against imperialism . . . in short, the great struggles for human emancipation are said to be over because they are, so the story goes, founded on what Haraway calls "the goad of human exceptionalism."[145] References to animals and to the environment in the discourses of transspecies posthumanism are on the one hand a consequence of the fact that capitalism does, in fact, pollute the environment and alienate and commodify the needs of all sentient beings, by subordinating all species to production for profit. Yet, the ideological abstraction of these problems from the capitalist mode of production—from class relations and private ownership of the means of production—is the means by which the ruling class ideologically sutures a seemingly "transclass" "care" for the planet and all of its creatures to its own class interests.

The "transspecies" or "green" family under capitalism is not an "evolutionary" break from capitalism or class relations but is an historical index of the expansion of capitalism into "green" products for the purpose of increasing profit. As global capitalism comes into crisis its class contradictions break through the surface and require new ideological legitimations that obscure the material source of the contradictions lying in the exploitation of labor in production. As the global economic crisis wears on, it has become routine for spokespersons of global capitalism to ideologically renew capital with a "green revolution," thus obscuring the exploitative relations which alienate the needs of the majority for the profit of the few. It is in this context that the discourses of transspecies posthumanism have not only come of age, but have received widespread institutional support. In the twenty-first century, the subordination of all aspects of life on the planet to private property and production for profit has reached new levels: from the private ownership of genes, seeds, salt, and entire species; to the industrialization and overproduction of wheat, milk, and animal slaughter, which is left to rot when it cannot be sold for a profit while people go starving; to e-waste and toxic computer dumping in exploited nations. Posthumanism represents the degradation of the environment as a cultural matter of "speciesism" that requires a new cultural revolution based on a transspecies ethics and thus obscures the class relations at the root of capitalism which actually explain the drive for profit and its destructive social and environmental consequences. By shifting the attention away from class exploitation to unethical speciesist practices posthumanism puts forward a vision of an ethical capitalism beyond capitalism, a capitalism that is above all sustainable and in which class exploitation is naturalized and taken to be the normal pattern of social organization.

Class as the Element of Life under Capitalism

Biopolitics in general and transspecies posthumanism in particular have become the norm in cultural theory because they provide a cultural updating of the social relations of production in capitalism that is in accord with new strategies in which capital seeks to realize surplus-value through "green" technologies. The greening of capitalism has become the new cultural horizon which justifies exploitation under the pretext of getting in touch with nature/the animal and the environment. More specifically, with the ideological "greening" of capitalist production, this means that the other which has been relegated to inhuman status—owing to the exploitation of the other's labor in production—is now discursively aligned with the animal and is revalued as a (re)source for a more just world or what Haraway calls "alter-globalization." This is a ruling class *ruse* for sentimentalizing exploitation. What this does is romanticize the poverty of the other which is not caused by speciesist ideology but by wage-labor/capital relations.

Transspecies posthumanism is ultimately a bypassing of the *social*—and the need for social transformation to abolish the exploitation of labor—and a return to a form of the "elemental" and the "natural." To put this another way, posthumanism ideologically normalizes the way in which wage-labor/capital relations, through the exploitation of surplus-labor and the alienation of workers from the social products of their collective labor, reduces what Marx calls "species life" into a "mere means to individual existence."[146] In *The Economic and Philosophic Manuscripts of 1844*, Marx argues for an important difference between "species life" and what he calls "natural" or "animal life."[147] "Natural life" is the life of "eating, drinking, procreating. . ." in short, of meeting immediate physical needs to reproduce individual life.[148] "Species-life," by contrast, is the life in which humans understand themselves in their world-historical, social relations and work to transform them. It is the life of conscious laboring activity in which humans understands themselves not simply as biological beings for individual reproduction and survival, but as social and historical beings who collectively produce and transform their social relations of production.

For Marx, *both* "natural life" and "species life" are *historical* which is to say that neither is outside of historical relations of production; both are mediated by the relations of production. For instance, *under capitalism*, even to maintain or reproduce "natural life," or "physical existence," Marx explains, an individual who does not own the means of production is compelled to maintain herself "as a worker" by selling her labor-power in order to acquire the means of subsistence to reproduce her individual physical existence.[149] As well, "species life" is enabled by the *historical* development of collective labor, of the forces of production. The difference, however, is that "natural life" is the *reproduction of individual life*, within the existing relations of production, whereas "species-life" is the conscious, collective, material transformation and production of social relations of production. Through the historical development of their collective labor, humans are "species-beings" who have not simply developed the capacity to meet and reproduce their immediate individual physical needs but have also developed the capacity to become conscious of and work to collectively transform their historical and social relations of production.

And yet, in capitalism, Marx argues, workers are "estranged" or "alienated" from their species-being. This is because in capitalism people are divided by material relations into two classes: those who own the means of production and therefore exploit the surplus-labor of others and make profits from it and the majority of others who own only their own labor and sell it for wages, which they pay back to the owners of labor to buy the food, medicine, houses, cars, etc., they need to go back to work for the owners of labor. In relations of production based on exploitation, the products of production are expropriated from workers and confront the worker as forces alien to them. This alienation of labor, moreover, is not simply an alienation of the worker from the specific products she produces but is an alienation within production relations themselves or, as Marx puts it, within "the act of production, within producing activity itself."[150] It is owing to the material relations of production, therefore, that "the worker [. . .] only feels herself outside of her work, and in her work feels outside of herself. She is at home when she is not working and when she is working she is not at home."[151]

It is owing to the *material relations* of production founded on private ownership of the means of production and the theft of surplus-labor that workers are alienated and that capitalism reduces the life of the species to "a mere means to individual existence."[152] Under capitalism, in other words, the reproduction of physical life within existing social relations of production becomes the main goal of life for the majority, while at the same time the minority who own the means of production, profit from the labor of the majority who only have their labor to sell to survive and reproduce themselves. Moreover, struggles to enable people to consciously grasp and direct their collective labor to transform their world by means of material praxis are marginalized. Capitalism, in short, reduces the majority "to work to live and live to work."

Biopolitics more generally, and transspecies posthumanism in particular, are theories of "passive adjustment" to the ruins of capitalism. They spiritualize poverty and the subordination of love, kinship, and sexual relations to commodity exchange relations and production for profit. They reduce species life to a mere means of individual survival within capitalism. This is a far cry from the understanding of "love" produced by historical materialists such as Kollontai who argued that the basis of the "hypocritical morality" of capitalism is not in its failure to produce "ideal (post)human beings"—what Haraway calls "companion species" or Hardt and Negri call "new and different subjectivities"—rather it is in its material relations of production. The hypocritical morality of capitalism is not an effect a specific kind of "love" or "family" (these are its *symptoms* and *articulations*) but rests on "the structure of its exploitative economy."[153] Freedom of sexuality, love, and desire cannot be produced unless emotional relations are, as Kollontai argues, "freed from financial considerations,"[154] which is to say, freed from class society and its privatized relations of production that produce dire economic necessity for the majority. This is not simply a matter of "meeting individual needs" for the reproduction of capitalism. Rather, it requires freedom from necessity. Freedom, that is, from social relations of production based on the exploitation of labor which, if left intact, will inevitably subordinate human relations including love and sexual relations to "financial considerations." "Love"—of animals, of people, of differences, of the world—does not evolve or transcend beyond capitalism without the material transformation of capitalist relations of production. Rather, it is in dialectical relation to the material relations of production in society. "Love" can only be "freed" if it is freed from class society and its privatized relations of production that subordinate the planet to production for profit while producing dire economic necessity for the majority. For an emancipatory theory of love what is needed is a return to grasping the class relations that structure life under capitalism and understanding that ending alienation requires bringing about social relations of production in which class antagonisms have not only already been abolished—because private property has been abolished—but have been, as Engels puts it, "forgotten in practical life."[155]

BIBLIOGRAPHY

Agamben, Giorgio. *The Coming Community*. Translated by Michael Hardt. Minneapolis: University of Minnesota Press, 1993.

---. *Homo Sacer: Sovereign Power and Bare Life*. Translated by Daniel Heller-Roazan. Stanford, CA: Stanford University Press, 1998.

---. *Means Without End*. Translated by Vincenzo Binetti and Cesare Casarino. Minneapolis: University of Minnesota Press, 1996.

Augustine. *Confessions*. Translated by R. S. Pine-Coffin. London: Penguin, 1961.

Barthes, Roland. *Mythologies*. Translated by Annette Lavers. New York: Farrar, Straus, and Giroux, 1972.

The Bhagavad-Gita. Translated by Barbara Stoler Miller. New York: Bantam Classics, 1986.

Cerni, Paula. "Imperialism in the Twenty-First Century." *Theory and Science* (2006). http://theoryandscience.icaap.org/content/vol8.1/cerni.html.

Corsani, Antonella. "Beyond the Myth of Woman: The Becoming-Transfeminist of (Post)Marxism." *SubStance* 112, 36, no. 1 (2007): 107–138.

Deleuze, Gilles, and Félix Guattari. *A Thousand Plateaus: Capitalism and Schizophrenia*. Translated by Brian Massumi. Minneapolis: University of Minnesota Press, 1987.

Eagleton, Terry. *Literary Theory: An Introduction*. 2nd ed. Malden, MA: Blackwell, 1996.

Ebert, Teresa and Mas'ud Zavarzadeh. *Class in Culture*. Boulder, CO: Paradigm Publishers, 2008.

Engels, Frederick. *Anti-Dühring*. Peking: Foreign Languages Press, 1976.

---. *The Dialectics of Nature*. Moscow: Progress Publishers, 1972.

Foucault, Michel. *The History of Sexuality*. Vol. 1, *An Introduction*. Translated by Robert Hurley. New York: Vintage, 1978.

Haraway, Donna. *When Species Meet*. Minneapolis: University of Minnesota Press, 2008.

Hardt, Michael, and Antonio Negri. *Commonwealth*. Cambridge, MA: Harvard University Press, 2009.

Kollontai, Alexandra. *Selected Writings*. Translated by Alix Holt. New York: Norton, 1977.

Lewontin, Richard and Richard Levins. *Biology Under the Influence*. New York: Monthly Review Press, 2007.

Marx, Karl, and Frederick Engels. *The German Ideology*. Moscow: Progress Publishers, 1976.

---. *Manifesto of the Communist Party*. In *Marx-Engels Collected Works*, vol. 6, 476–519. New York: International Publishers, 1976.

Marx, Karl. *Capital: A Critique of Political Economy*. Vol. 1. Translated by Ben Fowkes. London: Penguin Books, 1976.

---. "A Contribution to the Critique of Hegel's Philosophy of Right: Introduction." In *Critique of Hegel's "Philosophy of Right"*. Translated by Annette Jolin and Joseph O'Malley. Cambridge, MA: Cambridge University Press, 1970.

---. *Economic and Philosophic Manuscripts of 1844*. Translated by Martin Milligan. New York: International Publishers, 1964.

Plato. *Five Great Dialogues*. Translated by Benjamin Jowett. New York: Walter J. Black, 1942.

Roesch, Jen. "Turning Back the Clock? Women, Work, and the Family Today." *International Socialist Journal* 38 (Nov–Dec 2004). http://www.isreview.org/issues/38/women_family.shtml.

Wade, Nicholas. "New Finding Puts Origins of Dogs in the Middle East." *The New York Times,* March 17, 2010. http://www.nytimes.com/2010/03/18/science/18dogs.html?_r=0.

NOTES

1. Christian Marazzi as cited in Antonella Corsani, "Beyond the Myth of Woman: the Becoming-Transfeminist of (Post)Marxism," *SubStance* 112, 36, no. 1 (2007): 107.
2. Michel Foucault, *The History of Sexuality,* vol. 1, *An Introduction,* trans. Robert Hurley (New York: Vintage, 1978), 143.
3. Ibid., 140–41.
4. Ibid., 94.
5. Ibid., 93.
6. Karl Marx and Frederick Engels, *The German Ideology* (Moscow: Progress Publishers, 1976), 102.
7. Foucault, *History of Sexuality,* 159.
8. Ibid., 145.
9. Michael Hardt and Antonio Negri, *Commonwealth* (Cambridge, MA: Harvard University Press, 2009), 56–57.
10. Ibid.
11. Giorgio Agamben, *Homo Sacer: Sovereign Power and Bare Life,* trans. Daniel Heller-Roazan (Stanford, CA: Stanford University Press, 1998), 10.
12. Ibid., 9.
13. Ibid., 10–11.
14. Giorgio Agamben, *Means Without End,* trans. Vincenzo Binetti and Cesare Casarino (Minneapolis: University of Minnesota Press, 1996), 114–15.
15. Agamben, *Homo Sacer,* 82.
16. Giorgio Agamben, *The Coming Community,* trans. Michael Hardt (Minneapolis: University of Minnesota Press, 1993), 102.
17. Donna Haraway, *When Species Meet* (Minneapolis: University of Minnesota Press, 2008), 46.
18. Hardt and Negri, *Commonwealth,* 182.
19. Corsani, "Beyond the Myth," 107.
20. Ibid., 117.
21. Hardt and Negri, *Commonwealth,* 139–40.
22. Ibid., 140.
23. Paula Cerni, "Imperialism in the Twenty-First Century," *Theory and Science* (2006), under "The New Global Relations of Production," http://theoryandscience.icaap.org/content/vol8.1/cerni.html.
24. Ibid.
25. Karl Marx, "A Contribution to the Critique of Hegel's *Philosophy of Right*: Introduction," in *Critique of Hegel's "Philosophy of Right,"* trans. Annette Jolin and Joseph O'Malley (Cambridge, MA: Cambridge University Press, 1970), 131.
26. Ibid.
27. Ibid.
28. Ibid.
29. George Sampson as cited in Terry Eagleton, *Literary Theory: An Introduction,* 2nd ed. (Malden, MA: Blackwell, 1996), 21.
30. Alexandra Kollontai, *Selected Writings,* trans. Alix Holt (New York: Norton, 1977), 278.
31. Ibid.
32. Ibid., 278–79.

33. Ibid., 280.
34. Ibid.
35. Ibid., 280–81.
36. Ibid.
37. Ibid., 237–49.
38. Hardt and Negri, *Commonwealth*, 183.
39. Ibid., 56.
40. Ibid., 57 (emphasis added).
41. Ibid., 56.
42. Ibid., 57.
43. Ibid., 56–57.
44. Ibid., 57.
45. Marx and Engels, *German Ideology*, 42.
46. Frederick Engels, *The Dialectics of Nature* (Moscow: Progress Publishers, 1972), 170.
47. Karl Marx, *Capital: A Critique of Political Economy*, vol. 1, trans. Ben Fowkes (London: Penguin Books, 1976), 283.
48. Marx and Engels, *German Ideology*, 41.
49. Hardt and Negri, *Commonwealth*, 181.
50. Ibid., 180.
51. Ibid., 181.
52. Ibid., 180–83.
53. Ibid., 195.
54. Ibid., 192.
55. Ibid., 181–82.
56. Ibid., 193.
57. Ibid., 182–83.
58. Ibid., 183.
59. Karl Marx, *Manifesto of the Communist Party*, in *Marx-Engels Collected Works*, vol. 6 (New York: International Publishers, 1976), 487.
60. Hardt and Negri, *Commonwealth*, 195.
61. Ibid., 59.
62. Ibid., 194.
63. Augustine, *Confessions*, trans. R.S. Pine-Coffin (London: Penguin, 1961), 213.
64. Plato, *Five Great Dialogues*, trans. Benjamin Jowett (New York: Walter J. Black, 1942), 477–95.
65. Augustine, *Confessions*, 147.
66. Ibid., 148–49.
67. Ibid., 150–51.
68. Hardt and Negri, *Commonwealth*, 193.
69. Ibid., 198.
70. Ibid., 193.
71. Ibid., 193–94 (emphasis added).
72. Ibid., 129.
73. Ibid., 131.
74. Ibid., 141.
75. Ibid., 140.
76. Ibid., 191.
77. Ibid., 187.
78. Ibid., 186–87.
79. Ibid., 184–85.
80. Ibid., 185.
81. Ibid.
82. Ibid.
83. Ibid., 185–86.
84. Ibid., 186.

85. Ibid.
86. Ibid., 187–88.
87. Ibid., 198–99.
88. Félix Guattari as cited in Hardt and Negri, *Commonwealth*, 186.
89. Hardt and Negri, *Commonwealth*, 186.
90. . Ibid., 188.
91. Ibid., 186–87
92. Ibid., 188.
93. Ibid., 187.
94. Kollontai, *Selected Writings*, 297.
95. Ibid., 246.
96. Hardt and Negri, *Commonwealth*, 187.
97. Kollontai, *Selected Writings*, 263.
98. Ibid., 247.
99. Ibid.
100. Ibid., 237.
101. Jen Roesch, "Turning Back the Clock? Women, Work, and the Family Today," *International Socialist Journal* 38 (Nov–Dec 2004), http://www.isreview.org/issues/38/women_family.shtml.
102. Donna Haraway, *When Species Meet* (Minneapolis: University of Minnesota Press, 2008), 3–4.
103. Ibid., 46.
104. Ibid.
105. Ibid.
106. Marx, *Capital*, vol. 1, 283.
107. Nicholas Wade, "New Finding Puts Origins of Dogs in the Middle East," *The New York Times*, March 17, 2010, http://www.nytimes.com/2010/03/18/science/18dogs.html?_r=0.
108. Richard Lewontin and Richard Levins, *Biology Under the Influence* (New York: Monthly Review Press, 2007), 36.
109. Ibid., 37.
110. Haraway, *Species*, 3–4.
111. Ibid., 5.
112. Ibid.
113. Ibid.
114. Ibid., 3.
115. Ibid.
116. Ibid., 49.
117. Ibid., 51.
118. Ibid.
119. Ibid., 61.
120. Ibid., 97.
121. *The Bhagavad-Gita*, trans. Barbara Stoler Miller (New York: Bantam Classics, 1986), 69.
122. Ibid., 85.
123. Ibid., 91.
124. Ibid., 33.
125. Ibid.
126. Ibid.
127. Ibid., 34.
128. Ibid., 38.
129. Ibid., 61.
130. Ibid., 69.
131. Haraway, *Species*, 84.
132. Ibid., 73–82.

133. Roland Barthes, *Mythologies*, trans. Annette Lavers (New York: Farrar, Straus, and Giroux, 1972), 100–02.
134. Haraway, *Species*, 181.
135. Ibid., 69–70.
136. Nancy Farmer as quoted in Haraway, *Species*, 69.
137. Haraway, *Species*, 74.
138. Ibid., 72.
139. Ibid., 72–73.
140. Ibid., 73.
141. Ibid., 74.
142. Ibid., 69–93.
143. Ibid., 83.
144. Ibid.
145. Ibid., 46.
146. Karl Marx, *Economic and Philosophic Manuscripts of 1844*, trans. Martin Milligan (New York: International Publishers, 1964), 113.
147. See Teresa Ebert and Mas'ud Zavarzadeh, *Class in Culture* (Boulder: Paradigm Publishers, 2008), xi–xii, for a discussion of "natural life" and "species life" in Marx, *Economic and Philosophic Manuscripts of 1844*.
148. Marx, *Manuscripts*, 111.
149. Ibid., 109.
150. Ibid., 110.
151. Ibid.
152. Ibid., 113.
153. Kollontai, *Selected Writings*, 263.
154. Ibid., 274.
155. Frederick Engels, *Anti-Dühring* (Peking: Foreign Languages Press, 1976), 119.

TWO

Loving Transnationalism

Spiritualizing Class in House of Sand and Fog

Amrohini Sahay

The "spiritual turn" in cultural theory—from the later writings of Eagleton (*Culture and the Death of God*), Derrida (*Acts of Religion*), Badiou (*St. Paul: The Foundation of Universalism*), Žižek (*The Fragile Absolute*) . . . to its second-order appearance in such volumes as *Spiritual Shakespeares* as part of the Routledge Accent series which are central texts of dissemination of academic theory for large graduate and undergraduate audiences—has been much commented upon in recent years. What is missing in these discussions, however, is any concept that the latest turn in cultural theory from the material to the spectral is part of the formation of the ethical pedagogy of transnationalism to manage the contradictions of class relations now—by spiritualizing them. To be more clear, as the economic strength and global dominance of the advanced industrial economies of the West increasingly declines in an era of perpetual crisis, leading to both massive growth in inequality as well as "market opportunity," the culture industry, as the ideological arm of the transnational bourgeoisie, is ever more focused on inculcating subjects who respond to the new material conditions of labor in a way that defuses the class tensions and re-aligns subjects into a post-class "community": that is, in order to solve in the cultural imaginary what cannot be solved in the material economic relations. The turn to spirit in high theory as well as in its "everyday" counterpart, cultural studies, by way of emphasis on affect (whether theorized in terms of emotion and empathy or non-oppositional circuits of feelings and intensities which exceed the coordinates of the given categorical norms), is thus displacing the older rhetoric of an unbridled desire and the emphasis on the autonomous subject of difference "free" of any social constraints ("norms") which dominated cultural discourses since the 1960s.

In this historical context, the posthumanist turn is a subset of the broader spiritual turn and the wider conversion of the economic into the cultural in order to propose a spiritual (non-material) solution to the intensifying material contradictions of capitalism. Despite claims that central to posthumanism is a marked return to ontology and thus to a materialism that foregrounds as Cary Wolfe states "the embodiment and embeddedness of the human being in not just its biological but also its technological world, the prosthetic coevolution of the human animal with the technicity of tools and external archival mechanisms (such as language and culture),"[1] the posthumanist turn nonetheless remains spectral in dehistoricizing and dematerializing ontology. Ontology, in other words, may be represented by posthumanists as that which disrupts correlationist epistemology and moves theory closer to the study of "reality," but insofar as ontology is written as "a becoming governed by no necessity whatsoever" as Quentin Meillassoux asserts,[2] posthumanist ontologies erase the material conditions within which being dialectically develops and create a speculative "ideal" world in which the contradictions of *this* world are suspended, reducing the historicity of being to a contemporary variation on the idealist self-moving of the Platonic soul.[3]

In the guise of remaining open to the unforeseen possibilities of a noncausal ontology—a ruptural politics of the "event," as Alain Badiou defines it[4] —perhaps the most effective discourses of posthumanism draw on the new spiritual ideology of "common-ism" alongside the transspecies writings that posit the cause of all social hierarchies as lying in philosophical humanism. Donna Haraway, for instance, proposes that "[t]he discursive tie between the colonized, the enslaved, the noncitizen and the animal—all reduced to type, all Others to rational man, and all essential to his bright constitution—is at the heart of racism and flourishes, lethally, in the entrails of humanism."[5] On this account, as Marianne DeKoven elaborates, one of the main motivations for critical work on the animal/human divide has been "the parallels between animals and subjugated humans"[6] such that the elimination of this supposedly originary divide promises simultaneously a new means to postracial otherness in a shared world. Such a post-identity ethics of hybridization and mixture is in turn deemed more suitable to an age in which such "grand" projects of social change as class struggle against capitalism have been claimed as "outmoded,"[7] and in which, as Hardt and Negri formulate it, "[l]ove of the stranger, love of the farthest, and love of alterity"[8] functions as "an ontological event" which "marks a rupture with what exists and the creation of the new."[9]

So ideologically pervasive is the new spiritual common-ism that not only those sections of the left who have accommodated themselves to capital by rejecting the need for organized international class struggle and its requisite political form in the communist party, but even "new communists" like Jodi Dean who critique the more overtly ideological forms of cultural left theory to argue for a militant politics centered on the building of the communist party reproduce its ideological assumptions. Thus while for Dean the primary and urgent goal of the left is to build a party which can function as a locus of resistance to the entrepreneurial individualism which atomizes the social and forecloses on solidarity, nevertheless the collective political subject of the party needs to be understood not as Marx's proletariat, but simply "the people as the rest of us."[10] Hence building revolutionary collectivity requires, for Dean, the turn to an inclusive populism based on the model of the Occupy Movement and not on historical materialism. Commenting on Marx and Engels' materialist perspective which theorizes communism in terms of the "real movement which abolishes the present state of things,"[11] Dean argues that this "real movement" should be understood not in terms of the objective material conditions and relations which structurally negate private ownership as a viable historical property form (Marx and Engels' argument) but rather in terms of "*the expansion of voluntary cooperation.*"[12] But a "being-in-common" based on embodying a "will" to commonality and not on the founding contradictions of the objective structure of capitalism—which requires a vanguard party precisely as the theoretically advanced section of the international proletariat which grasps "the historical movement as a whole"[13]—is simply a popular frontism which denies the economic basis of the social and the political movement of the self-emancipation of the proletariat and articulates once again a "utopian, petty bourgeois socialism, i.e. the *dream* of petty bourgeois intellectuals."[14]

In thus privileging the "values" of an inclusive (even post- or trans-species) community (based on new affective relations of "love of the stranger" or a voluntarist "will" to unity), the emphasis in contemporary cultural theory *seems* to be a critique of and resistance to the relations of private property ownership and the individualism stemming from a market society which separates and divides workers globally and thus alienates them from collective existence. In fact, however, such value narratives need to be read as the latest strategy of capital to deground any grasping of the objective—the common position of the subject of labor in the relations of production—which could lead to a revolutionary collectivity capable of acting materially in its own interests. Instead, by positing a subjective "solution" to the emptying out of any common identity by commodity relations which produce workers as monadic/nomadic subjects-of-difference cut off and isolated from what Marx, in one of his richest early writings, calls "species-being,"[15] posthumanist common-ism in fact protects capital from any global critique and intervention.

To be more clear: as Marx points out in his discussion of alienation, the necessary "consequence of the fact that man is estranged from the product of his labour, from his life activity, from his species-being is the *estrangement of man* from *man*."[16] In such a historical situation of estranged labor, that is, persons are not only alienated from the founding conditions of collective existence ("species-life") and thus their material relations to others, but such alienation means that their collective life itself appears as only *a means* for individual life: the "good" of each individual is therefore pitted against the good of all the others. The response of the "new" posthumanist (transnational) ethics (like all forms of ethics) to this real, material social alienation and conflict is then of course an inverted one which follows the laws of an inverted society: it recognizes commonality only on the secondary plane of a cultural common-ness which attempts to suture over the trauma of the isolated subject of labor. Such recognition of the subject only as a non-objective (spiritual) being, and of collectivity only as an affective "bringing together" is of course, in the end, a relay of the religious world view—and in its current as in its older forms, while it responds to the real suffering and deprivation produced by an economic system based on private ownership of collectively produced wealth, should be seen as a device of class adjustment and reconciliation.

The "spirit-ing" of the new conditions of labor as a new affective commons is of course not limited to cultural theory (which is itself but a moment of an emergent mass cultural pedagogy). Contemporary films such as *House of Sand and Fog*[17] (based on the book of the same title by leftist writer Andre Dubus III, which became a major bestseller after Oprah, the arch pedagogue of the affective, ensured its mass circulation by placing it on her book list) thus do for a mainstream audience the identical ideological work that is done for academic audiences in a more technical and abstruse language. Contrary to liberal film critics who dismiss such films as morally trite and contrived, or to leftists for whom the representation of working class characters and their conditions of life are progressive in and of themselves, it is important to read such films carefully not only for how they frame the problems of class society but also the ways in which they provide exemplary lessons in the manufacturing of an "appropriate" class subjectivity in the contemporary property relations.

Formally, at the level of plot, *House of Sand and Fog* is the story of the battle for ownership of a modest suburban house in Northern California between its two central characters, Kathy Nicolo and Iranian-American immigrant and ex-colonel Massoud Behrani. As a result of a bureaucratic mix-up by the local state officials, Nicolo's inherited bungalow is placed on auction and she is evicted from her home. While she struggles to regain ownership of the house, it is bought for significantly less than its market value by the colonel who sees it as an opportunity to lift himself and his family out of poverty and back into an upper middle class life by "flipping" the house—that is, reselling it for several times the price at which he bought it. The conflict that then ensues between Behrani and Nicolo for ownership of the house is complicated not only by the fact that Behrani, having sunk his life savings into the property and quit one of his jobs, refuses to transfer the house back to Nicolo for anything less than the going market price, but by Nicolo's involvement with an emotionally needy cop, Lester Burden, who reinforces her racist perception of the Behranis. After the cop decides to take the situation into his own hands by threatening the Behranis with deportation, the ensuing sequence of events lead to a tragic end to the story as the death of Behrani's son drives him to end the lives of himself and his wife.

More fundamentally, by taking the house as the object of conflict between Nicolo and Behrani, the film metaphorically narrativizes—and attempts to address—the class crisis of the "American Dream" (of home ownership, stable consumption, and upward mobility) for workers on the ever wider margins of the global economy, both white and immigrant, who are denied the possibilities of a middle class life through no fault of their own. It does this, as I shall argue, on two interrelated levels: that of a reformulation of "acceptable" desire and that of an affective relation to the other as a utopian solution to their common economic situation as workers and laboring poor.

On the first level, Kathy Nicolo, recently divorced as a result of her desire for children (a desire which her husband did not share) and a recovering addict who works as a house cleaner in the homes of the rich, symbolically represents the position of a section of the white working class which no longer has access to the possibilities of an earlier generation of workers symbolized by her dead father. Positioned as a subject without the requisite "skills" necessary for middle class existence in a postindustrial economy—not only does she lack a college education but she lacks the cultural make-up ("motivation" and "savvy") of the successful worker—she is a worker who has fallen through the cracks of the global economy. (As she says to Lester Burden in telling him how it came about that she lost the house, "It took [my father] 30 years to pay it off. And it took me 8 months to fuck it up."[18]) Here the film both acknowledges the conservative argument that such a class fate is deserved and is an effect of the "laziness" and sense of entitlement of privileged American workers and, in accord with its "progressive" left-liberal ideology, undercuts it. Thus, on the one hand Nicolo's fate is framed by the colonel's comment that Americans "do not deserve what they have. They have the eyes of small children who are forever looking for the next source of distraction, entertainment, sweet taste in the mouth. We are not like them. We know rich opportunities when we see them." [19]Nevertheless, this view is denied by way of the colonels' own class fate. In his case, as the film shows, those workers who do possess the "drive" to succeed (such as Behrani himself, who not only possesses "cultural capital" from his previous life in Iran but is the "model immigrant" who works two menial jobs, keeps meticulous accounts and carefully tabulates even his most minor expenses), are also denied the American Dream. In doing so the film opens up the space for a critique of class-as-structure, which would show that class position is dependent not on the "merit" of individual workers but on the systematic and impersonal production of workers to fit (or not to fit) the various and historically changing needs of capital. But this is of course the very critique that it is ideologically unable to make and thus the "twist" is that the film ultimately argues in effect that such workers should be reconciled to their class fate: denied the "American Dream" ("the ownership society"), they should realize that it is an alienating dream anyway and should learn a new set of values which would help them live without it. Workers, the film says, should be happy with what they have.

At this first level, the ideological lesson that the film encodes is the disciplining of a "desire" which no longer corresponds to the realities of the economic world in which most American workers now live. "Excessive" desires—such as Nicolo's desire for children, Behrani's drive to reattain the class position that he had in the Shah's Iran, and even Lester Burden's dissatisfaction with his family life and desire for Nicolo—lead to disastrous situations for all of them. The pursuit of individual desire thus represents, within the moral economy of the film, an "overreaching" that dissolves the social connections necessary for well-being. Thus Nicolo should have done what she could to hold on to what her father's work made possible (the house) as well as to her husband, and it is because she forgot this that she lost both; Behrani should have remained content with having acquired American citizenship for himself and his family, and should have saved all of his earnings rather than keeping up the pretense of a lifestyle which he could no longer afford and thus jeopardizing his family and himself; and finally, Lester Burden should have been content with his wife, kids, and respectable (if low-level) job rather than becoming dissatisfied and destroying both in his pursuit of Nicolo. In this sense the film is a lesson in "shoulds" and a disciplining of "wants" which teaches the contemporary worker the "appropriate" relation to their material situation.

If the first moment of the film functions by way of negative example, it is in its second and related moment that it goes on to make its "positive" argument with respect to a call for new intersubjective relations based on affect and empathy as the solution to the alienating conditions within which workers find themselves. What ultimately drives the plot forward—and is reproduced formally at the level of the alternating points of view of Nicolo and Behrani, neither of which is privileged in the narrative—is that both have (within their own respective frames of reference) a "legitimate" claim to the house, and neither are, for most of the film, willing to see outside these frames of reference. Moreover, in the context of their conflict, neither the bureaucratic state nor the legal system can effectively adjudicate their claims or produce a mutually acceptable outcome for them. Cut off from one another both by their struggle for ownership of the house and by their racist and sexist readings of the "other," both thus engage in forms of violent relations with each other that have calamitous consequences beyond their intentions or control. And yet, by way of contrast to the tragic denouement, the film also provides for another potentiality of relations between the two. Thus toward the later moments of the film, after Behrani takes in Nicolo following her attempted suicide, we see both Behrani and Nicolo moving toward new and different ethical relations of openness to the other which, the film says, if only realized earlier could have formed the grounds for their reconciliation. The privileged frame for this new ethical relation established between the two by the end of the film is, quite predictably, the trope of family: but here the image of "family" as reworked through the networks of transnational migration. Thus although the film ends in tragedy, with the death of the racial others, it also provides the resolution of the conflict in the image of Nicolo lying in a fetal position between the dead bodies of Behrani and his wife—thus acknowledging their "common" relations as bound by a (neo-familial) ethics extending across the boundaries of race, nationality, and cultural difference. Indeed, this ethical relation to (the claims of) the other is only underscored when, in the final scene of the film, Nicolo symbolically rejects her claim to ownership of the house, and it is in this context that the pursuit of a "perverse" (white, masculinist, and nationalist) ethics by the cop functions as a negative counterpoint to the model of the new inclusive transnationalist ethics as post-white, post-masculinist, and post-nationalist.

What is of course telling here is that the grounds for Nicolo's and Behrani's movement toward recognition of their commonality despite their cultural differences is established in the narrative not by way of arriving at any new conceptual understanding of their common class position and interests, but simply by way of a new openness of "feeling" toward the other. As such, the film works (as does much of contemporary theoretical writing) primarily as a class lesson in the ethics of transnationalism which posits a relation to others based not simply on tolerance of difference (as in the old multicultural model) but on the notion of "the common"—an emphasis on "practices of interaction, care, and cohabitation in a common world"[20] which Negri and Hardt call "love,"[21] and which Donna Haraway elaborates in terms of a "worldliness and touch across difference."[22] Translated into the logic of the film, Nicolo and Behrani are brought together, that is, not as workers or by any knowledge that they have a commonality grounded in their identical material position in class society, but *affectively*—through the realization that the other can be an object of "care": can be mourned and loved rather than "hated." The new "common" reflexively established by the film is thus based not on the *shared material interests* of the structurally objective—which need to be grasped abstractly and conceptually—but on the subjective, the micrological plane of experience—and again, as in current cultural theory, it is in fact a deformation of class solidarity.

Far from aiding in the struggle against the violences of race and other hierarchical differences which culturally supplement capitalist exploitation, the planetary cosmopolitanism articulated in the new languages of posthumanist common-ism ideologically resolves the violence of capitalist hierarchies into a critique of discursive binaries and identity as such, and this collapsing of all distinctions—including the distinction of race as a global division of labor—transforms politics into an ethics which does the work of transnational, cross-border capital. It thus conceptually suspends the materiality of difference under capitalist social relations (which are maps of developing labor relations) and especially the difference of class (profit) under the guise of a new emancipatory "worlding" of the world. It is on the terms of this new "worlding" that the world's workers are urged, for example, to see their own powerlessness in production in terms of an ecstatic biopolitical productivity: "Economic statistics can grasp the condition of poverty in negative terms but not the forms of life, languages, movements, or capacities for innovation they generate."[23] Yet whether couched in the terms of such a relentlessly affirmative gaze (which redeems the "fallen" world of the poor), or by way of the discourses of mourning and melancholia (in such films as *House of Sand and Fog* and others which emphasize the brutality of the social operations of difference) what these narratives obviate is the need for internationally organized and collective proletarian resistance against the fundamental conditions of alienation and exploitation in the capital/wage-labor relation. In doing so, far from representing an innovative and new form of politics for the present they merely rehearse the spiritual songs of yesterday.

BIBLIOGRAPHY

Badiou, Alain. *Being and Event*. Translated by Oliver Feltham. London: Continuum, 2005.

---. *St. Paul: The Foundation of Universalism*. Translated by Ray Brassier. Stanford: Stanford University Press, 2003.

Dean, Jodi. "The Party and Communist Solidarity." *Rethinking Marxism: A Journal of Economics, Culture and Society* 27 , no. 3 (2015): 332–42.

DeKoven, Marianne. "Guest Column: Why Animals Now?" *PMLA* 124, no. 2 (2009): 361–69.

Derrida, Jacques. *Acts of Religion*. Translated and Introduction by Gil Anidjar. New York and London: Routledge, 2002.

Eagleton, Terry. *Culture and the Death of God*. New Haven, CT: Yale University Press, 2014.
Fernie, Ewan, ed. *Spiritual Shakespeares*. London: Routledge, 2005.
Gibson-Graham, J. K. *The End of Capitalism (As We Knew It): A Feminist Critique of Political Economy*. Oxford: Blackwell 1996.
Haraway, Donna. *When Species Meet*. Minneapolis: University of Minnesota Press, 2008.
Hardt, Michael, and Antonio Negri. *Commonwealth*. Cambridge, MA: Belknap of Harvard University Press, 2009.
---. *Multitude: War and Democracy in the Age of Empire*. New York: The Penguin Press, 2004.
House of Sand and Fog. Directed by Vadim Perelman. Universal City, CA: DreamWorks Home Entertainment, 2004. DVD.
Lenin, V. I. "Narodism and the Class of Wage-Workers." In *Lenin Collected Works*. Vol. 20, 105–08. Moscow: Progress Publishers, 1977.
Marx, Karl. *Economic and Philosophic Manuscripts of 1844*. In *Marx-Engels Collected Works*, vol. 3, 229–346. New York: International Publishers, 1975.
Marx, Karl and Frederick Engels. *The German Ideology*. In *Marx-Engels Collected Works*, vol. 5, 19–584. New York: International Publishers, 1976.
---. *Manifesto of the Communist Party*. In *Marx-Engels Collected Works*, vol. 6, 477–519. New York: International Publishers, 1976.
Meillassoux, Quentin, "Potentiality and Virtuality." Translated by Robin Mackay. In *The Speculative Turn: Continental Materialism and Realism*, edited by Levi R. Bryant, Nick Srnicek, and Graham Harman, 224–36. Melbourne, Australia: re.press, 2011.
Plato. *Phaedrus*. Translated by Robin Waterfield. Oxford: Oxford University Press, 2002.
Wolfe, Cary. *What Is Posthumanism?* Minneapolis: University of Minnesota Press, 2010.
Žižek, Slavoj. *The Fragile Absolute: or Why is the Christian Legacy Worth Fighting For?* London: Verso, 1999.

NOTES

1. Cary Wolfe, *What Is Posthumanism?* (Minneapolis: University of Minnesota Press, 2010), xv.
2. Quentin Meillassoux, "Potentiality and Virtuality," trans. Robin Mackay, in *The Speculative Turn: Continental Materialism and Realism*, eds. Levi R. Bryant, et al. (Melbourne, Australia: re.press, 2011), 226.
3. Plato, *Phaedrus*, trans. Robin Waterfield (Oxford: Oxford University Press, 2002), 27–28.
4. Alain Badiou, *Being and Event*, trans. Oliver Feltham (London: Continuum, 2005), xii.
5. Donna Haraway, *When Species Meet* (Minneapolis: University of Minnesota Press, 2008), 18.
6. Marianne DeKoven, "Guest Column: Why Animals Now?" *PMLA* 124, no. 2 (2009): 363.
7. J. K. Gibson-Graham, *The End of Capitalism (As We Knew It): A Feminist Critique of Political Economy* (Oxford: Blackwell 1996), 112.
8. Michael Hardt and Antonio Negri, *Commonwealth* (Cambridge, MA: Belknap of Harvard University Press, 2009), 188.
9. Ibid., 181.
10. Jodi Dean, "The Party and Communist Solidarity," *Rethinking Marxism: A Journal of Economics, Culture and Society* 27 , no. 3 (2015): 339.
11. Karl Marx and Frederick Engels, *The German Ideology*, in *Marx-Engels Collected Works*, vol. 5 (New York: International Publishers, 1976), 49.
12. Dean, "The Party," 338.

13. Karl Marx and Frederick Engels, *Manifesto of the Communist Party*, in *Marx-Engels Collected Works*, vol. 6 (New York: International Publishers, 1976), 494.

14. V. I. Lenin, "Narodism and the Class of Wage-Workers," in *Lenin Collected Works*, vol. 20 (Moscow: Progress Publishers, 1977), 106.

15. Karl Marx, *Economic and Philosophic Manuscripts of 1844*, in *Marx-Engels Collected Works*, vol. 3 (New York: International Publishers, 1975), 275–77.

16. Ibid., 277.

17. *House of Sand and Fog*, directed by Vadim Perelman (Universal City, CA: DreamWorks Home Entertainment, 2004), DVD.

18. Ibid.

19. Ibid.

20. Hardt and Negri, *Commonwealth*, viii.

21. Michael Hardt and Antonio Negri, *Multitude: War and Democracy in the Age of Empire* (New York: The Penguin Press, 2004), 12.

22. Haraway, *When Species Meet*, 14.

23. Hardt and Negri, *Commonwealth*, xi.

II

New Materialisms, Object Ontologies, and Class Totalities

THREE

"Theory Too Becomes A Material Force"

Militant Materialism or Messianic Matterism?

Stephen Tumino

The "Wholly Other" Mystification of Labor

Currently a "material turn" is underway in the humanities away from the dogma of textual immanence and the "materiality of the signifier"[1] in the attempt to address the growing inequalities forming in the aftermath of the crisis of global capitalism. And yet, what is taken to be "material" in the new materialisms—whether it be Derrida's "wholly other,"[2] or Deleuze's vitalist "immanence,"[3] Hardt and Negri's "multitude,"[4] or Žižek's "materiality of ideology,"[5] Haraway's "affective consubstantiality,"[6] or the "autonomy of the nonhuman"[7] and, the "vibrant"[8] "agency of matter"[9]—conforms to Derrida's call for a "materiality without materialism and even perhaps without matter," that is, the material as that which "exceeds" and "resists" the conceptual.[10] What is material for Derrida is the priority of alterity within iteration that displaces explanation of the actual. Derrida makes "matter" and the undoing of "materiality" in its general sense as movement of the "outside" independent of the subject, the focus of any inquiry on materialism. It is, of course, this role of matter as a spectral agency that disrupts and eludes the subject that resonates with the newly dominant formations of posthumanist[11] theory for which "the assertion that the human species is the dominant force shaping our world [. . .] fits all too easily with the modernist assumption of human mastery over nature [. . .] which has brought about our environmental crisis."[12] However, this posthumanist view of the material as exceeding conceptual mastery is put in question by historical materialist thinkers who read the deconstruction of materialism as a form of "matterism."[13] Matterism ultimately treats matter as that which resists human thinking and control and which, in effect, ends up substituting pan-physicalism for the analysis of the historical conditions shaping the human and nonhuman world under capitalism. For historical materialism, materialism is not matterism; it is not the primacy of matter that is at issue (undoing matter as Derrida does simply confirms its primacy, it does not suspend it). Materialism, on the contrary, is, as Fredric Jameson explains, "the ultimate determination by the mode of production."[14] On such a basis the ethical aura that the spectral agency of the inhuman other has taken on in the posthumanities represents a reification of class interests that benefits the ruling class by occulting the exploitation of labor that is central to capitalism. My argument here is that the classical Marxist theory of the material as the movement of the mode of production has been entirely abandoned in the dominant cultural theory not only because the explanatory knowledge it provides exposes the relations of exploitation on which disaster capitalism depends, but because it reveals these relations to be in the end only transformable by the agency of labor rather than a change in values or a new ethics.

Taking the centrality of Derrida's undoing of materiality as my starting point, I will focus on the question of the materiality of the new materialism in cultural theory, paying especially close attention to the claim offered by Pheng Cheah that Derrida's "non-dialectical materialism"[15] offers a more "fundamental"[16] and "systemic"[17] understanding of materialism than "Marx's understanding of material existence"[18] as shaped by labor, which for Marx is the life activity of the human species. According to Cheah, because Marx's dialectical materialism is premised on "labor as a process [. . .] whereby given reality [. . .] is negated"[19] and "the radical transformation of existing social relations"[20] effected, it exhibits the "subordination of potentiality to actuality"[21] typical of humanism which always conceives matter as "negated through the imposition of a purposive form."[22] Conversely, because Derrida's "non-dialectical materialism" makes the material a "weak [messianic] force"[23] that "resists [. . .] any purposive or end oriented action [. . .] based on rational calculations"[24] it represents a new, more fundamental form of materialism, according to Cheah. At stake in Cheah's opposition of non-dialectical to dialectical materialism, therefore, is the question of the radical today: Is it "radical," as Marx argues, to "grasp things by the root"[25] by placing them within "the ensemble of the social relations"[26] (what Lenin called "militant materialism"[27])? Or is radical now, as Derrida claims, "a materialism without substance: a materialism of the khôra for a despairing 'messianism'"[28] —that is, a messianic materialism that mystifies the social as a chaotic flux of cause-less arrivals and spectral events that undoes the positive and reliable knowledge of the class totality that workers need for their emancipation from capitalist exploitation?

What makes Cheah's essay, "Non-Dialectical Materialism," of interest is in part its inclusion in the recent *New Materialisms* anthology, a text which is symptomatic of the current "turn to materialism" more generally. Its inclusion there would seem to call into question the claim made by the books' editors in the introduction that the "textual approaches associated with the cultural turn are increasingly being deemed inadequate for understanding contemporary society"[29] because of their failure in "thinking about matter, materiality, and politics" as shaped by the "contemporary [. . .] global political economy."[30] After all, Cheah claims in this essay that Derrida's "figure of the text in general"[31] represents the most "fundamental"[32] and "systemic"[33] understanding of materialism to date and argues against the transformative force of labor or indeed any concept of the material as a force of negation that might lead to theory becoming a guide for social praxis. Leaving aside Cheah's essentialist and, therefore, self-negating defense of Derridean tropes as most "fundamental" and "systemic" and how this reveals that the issue is not which theories of the material are "metaphysical" and which are not, as he claims, he is certainly quick to distance his defense of Derrida's textual materialism from the taint of De Manian "literariness," arguing that it is a "mistake" to take the "materialist understanding of text as [. . .] a self-interiority without an outside."[34] Rather, he understands the textual in Derrida's writings to be a defense of materialism as a "philosophy of the outside"[35] opposed to all "metaphysics" that deny the "force of materiality,"[36] which he takes to be "a limitless weave of forces or an endless process or movement of referral"[37] instantiated by our "contemporary technomediated reality."[38]

And yet the equation of the material with the virtual real shows that what Cheah takes to be material is not the "outside" (the class structure) but rather the "inside" (the cultural superstructure). What is material on this view is how techno-culture reveals the power of a virtually "inappropriable other" over the actual such as to normalize an "experience of an incalculable justice that escapes all rule,"[39] or, in other words, a sense of "urgency" that "forces us to act" without reason.[40] For Cheah, the "force of materiality" is "nothing other than the constitutive exposure of (the subject of) power to the other,"[41] which, citing Derrida, means "the absolute other in me [. . .] that decides on me in me."[42] In other words, the material is an opacity that defies understanding, as in Kant's view of the noumenal thing-in-itself which can only be known at the level of its phenomenal effects in consciousness; however, in Cheah's view, rather than being made available to reason, as in Enlightenment thought, it can only be experienced from within as an emotional plenitude that eludes analysis or explanation and consequently makes individuals feel weak and vulnerable. Not only does this translate the use of technology under capitalism to increase the rate of accumulation of surplus value into the autonomous force of technology itself, to which the subject must learn to willingly submit herself, it also turns the material into the intensity of feeling (weak in the face of an overwhelming and anonymous power) rather than the determination of the cause of this feeling of disempowerment in the social totality. This is significant for two reasons. On the one hand, by treating technology as both inappropriable and as that which forces humans to act (on their beliefs), Cheah of course re-situates the technological non-human not as passive and inert relative to the human but as possessing its own agency and ability to act on and control humans. On the other hand, such an experience with the non-human on the part of the human is equated with an unthinking compulsion that is usually disavowed in humanist thought as proper only to animals and machines. It is the sheer affirmation of such an experience of impersonal power that is taken to be the limit of the material in posthumanist discourses and made to seem inherently ethical. Despite differences of idiom, this is the dominant presupposition underlying all of the new materialisms today, which privilege, in the words of Jane Bennett, the material as "an excess that escapes quantification, prediction and control,"[43] and which claim this as the basis of ethical action because it reveals the co-dependency of the human with the non-human and inhuman. What is elided, however, by this affirmative matterism which contests the domination over nature, is the question of the function of the ethical to obscure and thus maintain the systemic class inequality inscribed in the daily exploitation of labor; but this is what in the end alone explains the capitalist mis-use of nature for profit.

It turns out that what Cheah means by "non-dialectical materialism" is a kind of technological determinism in which technology functions not as the social mediation between humans and the natural world under specific historical conditions (class relations) but, rather, aesthetically, as what inscribes bodies with moving experiences that he maintains are always to be ethically affirmed so as to change "the very idea of political organization,"[44] especially "in terms of creative labor qua negativity [. . .] embodied in the proletariat as a sociohistorical subject."[45] Marx's dialectical materialism of course requires that the apparent singularities of our experience, what Marx calls the "imagined concrete,"[46] be conceptually grasped as "the concentration of many determinations, hence unity of the diverse"[47] so as not to be deceived by such appearances into assuming a "chaotic conception of the whole."[48] Cheah, following Derrida, argues for "overturning [organization] as [. . .] the central principle of dialectical materialism,"[49] because of how it assumes that "the dynamism of matter comes from the activity or process of [. . .] ordering things through [. . .] relations of interdependence such that they become parts or members of [. . .] an integrated or systemic totality."[50] In Derrida's terms, the overturning of "organization" (and therefore dialectical materialism) requires deconstruction of the binary of "organic and [. . .] inorganic" as they represent for him the "two predicates that are most often attributed without hesitation to matter or to the material body" that "carry an obvious reference [. . .] to the possibility of an internal principle that is proper and totalizing" of both, which is "precisely, organization."[51] It is in order to undo the dialectical materialist concept of "totality" that alone explains why the organization of labor under capitalism gives rise to the proletariat as a revolutionary class, that Derrida has argued for the necessity of a "machinistic materiality without materialism and even perhaps without matter."[52] According to this narrative, it is within "machinistic materiality," or a "new figure of an event-machine,"[53] that the binary of the organic and inorganic is dehierarchized such that technology will no longer be thought as tools impassively receiving commands in a "state of anesthesia [. . .] without affect or auto-affection, like an indifferent automaton"[54] but will instead articulate "events of a kind that ought to resist any mechanization, any economy of the machine, namely [. . .] acts of [. . .] faith."[55] Cheah clarifies that what is at stake in Derrida's call for "a certain materiality, which is not necessarily a corporeality [but] a certain technicity"[56] is the kind of "creative appropriation" of technology found in the writings of Hardt and Negri, which attempt to construct "a sociohistorical subject that replaces the proletariat in contemporary globalization."[57] For Cheah this is the subject open to the "experience of an incalculable justice that escapes all rule"[58] which acts as a "weak [messianic] force" in individuals that defies "any purposive or end-oriented action [. . .] based on rational calculations or the projection of an ideal end"[59]—such as the end of class exploitation and the emancipation of the proletariat. The subject is thus reified from its insertion in class relations and taken to be a ghost in the machine or a body without organs that spontaneously resists explanation in terms of a causal material outside. In short, it is the subject as imagined in bourgeois ideology that is required to naturalize the exploitation of the workers as a voluntary act rather than as the necessary consequence of the means of production being owned as private property.

Cheah rejects Marx's dialectical materialism premised on the agency of labor by turning labor into a non-dialectical trope for "the vital body of the organism"[60] whose agency consists in the negation of matter through "the imposition of purposive form."[61] Implicit here of course is the familiar (de Manian) poststructuralist critique of romantic organicism—an argument which is used to reject such Marxist categories as labor and class for being essentialist and homogenizing, so as to oppose Marxism on epistemological grounds as a "metaphysical [. . .] humanism of the hand,"[62] for instance. But the point I want to emphasize here is that labor for Marx is not a "thing," such as the body or the physical activity of individuals. Although Marx makes labor the "essence of man" he clarifies that this is not in the sense of "an abstraction inherent in each single individual,"[63] as the purposive actor found in Aristotle's writings for example, or the conception of "society as the subject" found in Hegel, for instance.[64] For Marx, labor is "the ensemble of social relations."[65] It is this that distinguishes humans from animals:

> Men can be distinguished from animals by consciousness, by religion or anything else you like. They themselves begin to distinguish themselves from animals as soon as they begin to produce their means of subsistence [. . .] This mode of production must not be considered simply as being the reproduction of the physical existence of the individuals. Rather it is a definite form of activity of these individuals, a definite form of expressing their life, a definite mode of life on their part. What they are, therefore, coincides with their production.[66]

Marx contrasts labor with both the purposive concrete activity of individuals (what he calls "work"), the *homo-economicus* of bourgeois political economy, as well as the animal life-activity required by all species for their immediate physical survival. Cheah's failure to understand Marx's labor theory, which attributes to Marx the very bourgeois ideology he in fact opposes, is not simply a cognitive failure on his part, however, because it represents a mystification of labor as physical "work" that is typical of "the fetishism which attaches itself to the products of labour as soon as they are produced as commodities"[67] that Marx's labor theory critiques so as to produce a materialist understanding of the social totality. The fetish of the vital body that negates matter that Cheah attributes to Marx is in actuality the object of Marx's critique of bourgeois political economy, which, as he demonstrates, fails in theory to get beyond the limits imposed by the capitalist mode of production in practice. The fetish of labor as concrete physical work in bourgeois theory is an ideological reflection of the actual reduction of the life-activity of human beings to animal life activity undertaken for mere survival (wage-labor).

Marx of course recognizes that, "the life of the species, both in man and in animals, consists physically in the fact that man (like the animal) lives on organic nature."[68] This physical life-activity or reproductive activity is what for Marx defines a "species-being" in general. The difference between the life-activity of the human species (labor) from that of other animals, however, is that it produces a surplus over and above what is required for immediate physical existence and in the process transforms the environment, and thereby, not only transforms humanity but also all life on Earth. It is this all-round transformative life activity of the human species that Marx calls "labor" in distinction to "work." Finally, it is because of labor that humans can in turn be distinguished from other animals by their "consciousness":

> The animal is immediately one with its life activity. It does not distinguish itself from it. It is its life activity. Man makes his life activity itself the object of his will and of his consciousness. He has conscious life

> activity. It is not a determination with which he directly merges. Conscious life activity distinguishes man immediately from animal life activity.[69]

The "wholly other" that arises out of techno-mediated culture which Cheah thinks constitutes an ethical subjectivity in the daily that disrupts Marx's theory of labor as an organizational basis for social change is simply a mystification of the way human life-activity is dominated by capital and undertaken merely to profit a few over meeting people's all-round needs, including the need to live in a safe and healthy environment. In this way, by only seeing in labor the alienated capitalist form of it (work), Cheah naturalizes the exploitation of labor at the center of capitalism. In doing so he cannot accept the dialectical self-negation of the working class as a class-in-itself—that is required to submit to being exploited in order to live—into a revolutionary class-for-itself—that must of material necessity become conscious of its alienation from its own life-activity and thereby undertakes to emancipate itself from the regime of wage-labor. He therefore mystifies agency as a mysterious movement of the "wholly other" and thus helps maintain bourgeois rule.

The Spiritual Fix Is In

As I have already mentioned, the *New Materialisms* anthology is symptomatic of the class conflicts which are driving the talking about (as an "other") and the untalking (as labor) of materialism today. In their introduction to what they are calling "new materialisms," Coole and Frost make what they say is a "somewhat arbitrary" division of materialism around three topoi: ontology, agency, and politics. They argue that all three "converge" on the same fundamental "point" that a more "ethical"[70] theory is needed now that the "textual approaches associated with the cultural turn are increasingly being deemed inadequate for understanding contemporary society."[71] Why cultural theory (which the editors equate with poststructuralism) is "inadequate" for understanding the contemporary is therefore a central question that frames the essays in the book. Yet dividing materialism today by way of the "arbitrary" presumes, of course, that such divisions over materialism are simply a discursive matter, as if knowledge makes the world, rather than providing a materialist explanation of how all discussions of materialism reflect the ongoing class conflicts over material resources that structure the daily. Such an "immanent" approach to materialism, far from being a "new" more "ethical" theory, is in fact the dominant cultural theory in transnational capitalism because it suppresses awareness of an "other" materialism that explains the material as the "ultimate determination by the mode of production."[72] What the publication of the book thus shows is how culturalist theory is currently being updated by co-opting the language of materialism so as to accommodate itself to capitalism in crisis by marginalizing class consciousness in the knowledge industry.

In *New Materialisms* theory is said to be inadequate in the contemporary because it is "associated with a cultural turn that privileges language, discourse, culture, and values"[73] and fails to "give material factors their due in shaping society and circumscribing human prospects."[74] It seems therefore that what is being proposed is the need for theory to become more reflexive about materiality by investigating the material conditions that shape culture, or as Coole and Frost put it, by "thinking about matter, materiality, and politics" as shaped by the "contemporary context of biopolitics and global political economy."[75] Leaving aside for the moment the supplementing of political economy with biopolitics and the way this treats power as separable from class in the manner of Foucault, when they come to describe these "material factors" that have placed cultural theory in crisis, rather than provide the systemic analysis of the "global political economy" they say is needed, what Coole and Frost instead do is list what has become a familiar litany of current events that have been popularized in the writings of Badiou, Žižek, Hardt and Negri among others as "problems of the common": "issues such as climate change [. . .] global capital and population flows, the biotechnological engineering of genetically modified organisms, or the saturation of our intimate and physical lives by digital, wireless, and virtual technologies."[76] In place of a systematic consideration of the contradictions of global capitalism that explains why what appear to be isolated issues (climate change, biogenetics, global inequality, etc.) are rooted in the law of value that rules the system of capital/wage labor relations, they list a series of cultural disruptions which, like all events, appear as causeless arrivals that elude explanation and that require their own immanent understanding, thus the pluralization of materialism into ontology, agency, and politics that frames the book. It turns out that the "somewhat arbitrary" dispersal of materialism into ontology, agency, and politics announced in the title is meant to address these crises under the assumption that they herald fundamental changes in the material world that cannot be explained in terms of the laws of motion of capitalism to be found in classical Marxism. The effect of giving a list of topics in place of a systemic understanding of capitalism is to segregate the social into localities and block the kind of materialist analysis that the editors themselves argue is above all necessary to address the impasse of the cultural turn.

The translation of the contemporary crisis of capitalism into the "problems of the common" distracts attention away from a causal understanding of the social totality in terms of the ongoing global class struggles over material resources. Contrary to the "new materialist" doctrine, however, why cultural theory is in crisis is not because of its own purely immanent epistemic blindness toward current events, but because the forces of production have advanced beyond the state capitalism of the 1970s (which in cultural theory was regarded as "totalitarian") as well as beyond the deregulated capitalism of the 1980s and 1990s (which was represented as "radical democracy" in cultural theory), and poststructuralist knowledges no longer have the power to normalize the social relations of global capitalism in crisis. New "materialisms" are thus necessary to reconcile in the cultural imaginary the ways in which the newer forces of production are calling into question the private ownership and control over the social wealth by capital at the expense of the workers, as is demonstrated by the economic crisis which daily exposes the immense accumulation of wealth which is unable to be put to "productive" (i.e., profitable to capitalists) use, while the conditions of life for workers globally deteriorates. The mode this accommodation to the dominant ideology takes is an "affirmative" materialism in defense of the "common" that demonizes sharp conceptualizations of the issues and critique for social change as "unethical." By replacing explanation of the emerging class conflicts with the "ethical" acknowledgement of them contra the blindness of discourse theory, the result is that the failure of capitalism is presented as a "crisis of values" that requires spiritual fixes.[77]

In William E. Connolly's essay, "Materialities of Experience," this approach of resignifying the economic crisis as a cultural crisis of values is given its proper religious tone when he argues for the force of "dissonances" in daily life—such as "a stock market crash disrupt[ing] assumptions about the future" or a "terrorist attack [that] folds an implacable desire for revenge in you"—to "disrupt[] conventional habits of perception"[78] and effect a "spiritual awakening" that will "cultivate further strains of attachment to this world and [. . .] fend off the seed of abstract resentment."[79] By fetishizing the affective experience of living in the crisis of global capitalism, rather than explain the experience in terms of its underlying cause in property arrangements, what is being opposed to the political economy of capital is an "affective political economy of loss and melancholia at the heart of the subject"[80] that the contributors to *The New Materialisms* all in their own way propose to heal with the ethos of the common. It turns out that the supplementing of political economy with biopolitics that Coole and Frost propose in their introduction is a way to say that the "new" materialism uses a Spinozan ethics of embodiment that imprisons the subject in her own experience of the crisis in order to block any understanding of its roots in terms of the systemic contradictions of capitalism.

On such experiential terms, class, rather than being an explanatory "struggle concept"[81] that lays bare the root conflict over property relations so as to change them, is instead taken to be a feeling of alienation that needs to be overcome by people becoming more ethical. Class is a "resentment" that comes from a sense of "entitlement" in "objectivity" that is held to be impossible,[82] or it is a "metaphysics" of "opposition" that is inhospitable to the other,[83] or a "gloomy and pessimistic"[84] "Thanatos-politics"[85] that refuses to accept and "move beyond"[86] the "impossibility of adequate compensation."[87] Even Jason Edwards' essay, "The Materialism of Historical Materialism," which critiques Hardt and Negri for obscuring "the character of contemporary capitalist societies" with "the ontological myth of the multitude"[88] and defends the need "to remember the materialism of historical materialism"[89] for its explanation of how "social production and consumption"[90] or "material life is organized, disrupted, and transformed,"[91] says that Marx's "notion of a 'class in itself' has to be jettisoned" as an unrealistic fantasy because class is "the experience of lived space."[92] Class, in other words, is a shared feeling in common that is endlessly (re)constituted in the micro-practices of everyday life rather than the exploitation inscribed in the work day. Disconnected from the underlying exploitation of labor by capital inscribed in the material base, class is unavailable as "an agent of social and political transformation," he concludes.[93] It seems that Lenin was right that under the bourgeois law of value in which science is used to increase profits, accumulate capital, and deepen inequality, "materialism" in Left and "ethical socialist" discourses is equated with "helping the exploiters to replace the old, decayed religious superstitions by new, more odious and vile superstitions" that produce more "ideological slaves of the bourgeoisie."[94]

The refusal to conceptualize the contemporary crisis of capitalism in terms of its "base"-ic class arrangements by remaining at the level of its surface effects where they appear as "problems of the common" has the effect of dispersing the social into a series of disruptions that are made into occasions to raise ethical choices in the daily and make the material, as in the pun, seem like what emotionally "matters" to people (beliefs, values, ethics). Thus, even when Coole and Frost articulate what they consider the concrete reasons for their conclusion that cultural theory is outdated, the way in which they do so makes the material seem "immaterial": a question of cultural disruptions, shifting knowledge, values, and interpretations, rather than systemic causes rooted in class arrangements. This translation of the material into the immaterial is also clear in the way they displace explanation with description in their account of the cultural turn as due, they claim, to "the exhaustion of once popular materialist approaches [. . .] such as [. . .] structural Marxism [. . .] since the 1970s."[95] In other words, people began turning to "culture" because Marxism was boring. In a similar way, the "new materialism" is "explained" as due to "the eclipse of materialism in recent theory."[96] In this way a description of a change in ideas (the spectral arrival of the "new") is put in place of a materialist explanation. And yet again when they discuss the "new materialist writing" and rather than inquire into the material conditions that explain why now a vitalist language of "forces, energies, and intensities" that represents the social as "random processes" that resist causal explanations is taken to be the limit text of materialism in the dominant knowledges they instead affirm how these "tropes and rhythms" have consolidated a "new currency" in the humanities that "might reconfigure our models of society and the political."[97] By equating linguistic change with social change they are assuming, as in bourgeois ideology, that knowledge makes the social rather than the reverse. It seems "odd" then that they say they are opposing the "textual approaches associated with the cultural turn" as "inadequate"[98] when they arrive at the same intellectual and political conclusions as the dominant theory that claims discourse shapes the world. What such an "event"-ful reading of the contemporary as constituted by epistemic shifts does is construct the "new" (materialism) as a break with the "past" (culturalism) so as to stage an occasion to revive the old materialism (matterism, vitalism) and make it seem like an ethical alternative that can save capitalism in crisis rather than what it is—an accommodation to the dominant ideology which is everywhere being delegitimated because of the growing inequality caused by the global exploitation of labor by capital.

It is because the new materialisms share the same fundamental assumptions about the material as culturalist theory that the editors can include Cheah's essay in their book—a text which reiterates the dogma that deconstruction is the most "radical"[99] materialism because it produces a "more fundamental"[100] and "systemic"[101] understanding of materialism than "Marx's understanding of material existence" premised on the "negative dialectic" of "creative labor"[102]—despite their claim that the "cultural turn that privileges language, discourse, culture, and values"[103] has been intellectually and politically delegitimated. Indeed, when Cheah claims that Derrida's "materialism without matter" has nothing to do with "the 'free play' of textual indeterminacy" of literary criticism,[104] he exhibits the same defensiveness regarding culturalist theory that Coole and Frost do when they impugn the ethics of the "cultural turn." And yet, the concept of materialism that is being defended in *New Materialisms* is precisely one that makes the material into an opacity that undoes causal explanation and can only be "interpreted" at the level of signs. The emphasis on interpretation over explanation is then taken as a mark of ethical belonging in the new common beyond class. What is held in common in the common, it turns out, is the commitment to "break with the concept of negation" by burying ideology critique and thereby show the kind of "hospitality to the other" that is imperative "in order to act in a responsible manner."[105] Critique is "negation" of the common because it brings to bear the disavowed other of class that critiques bourgeois ideology to open space for revolutionary change. It is this opening to the new—when theory becomes militant by gripping the masses with root knowledge—that "new materialisms" consider "unethical" because it violates the "real."

Critique Is the Matter of Class

The reunderstanding of materialism in the *New Materialisms* anthology as an "ethical turn" after poststructuralism is hardly a new materialist thinking. What exactly is being called "material" here if its status as material is that it "revalues" what was "devalued" in the past (culturalist) Theory? Furthermore, what is so "ethical" about reinscribing the othering of historical materialism? Rather than provide a materialist explanation of why everyone is suddenly a materialist now and so quick to distance themselves from the taint of culturalism, *New Materialisms* gives a culturalist interpretation of materialism as an ethical turn in the dominant ideas in response to cultural disruptions. Materialism, however, as Cheah cites Derrida in his essay, is always a "philosophy of the outside."[106] By constructing materialism as an "ethical turn" in the dominant knowledges that now recognizes Theory's other(s), Coole and Frost reinscribe the outside in terms of the inside so as to contain what Marx called the "material force" of theory as critique when it has "gripped the masses" by "grasp[ing] the root of the matter."[107]

Culturalism—the theory that what is material are signs, discourses, values, beliefs—has, contrary to the editors' narrative, been delegitimated not because it is now taken to be unethical for being insufficiently materialist. It is materialist in exactly the same way as the new materialisms are said to be because, in place of providing positive and reliable knowledge of the conflict of capital/wage labor relations "outside" discourse, it turns materialism in on itself and represents such a move (immanence) as an ethical calling that bears witness to the "other" as that which "exceeds" explanation and disrupts the conceptual. Because the material is an "absolute other"[108] that eludes representation and can only be known at the level of its "affects," materialism becomes thick description of the surfaces of culture rather than knowledge of the outside. This immanent approach to materialism reverses cause and effect as in Nietzsche's writings where "cause" is impossible to grasp outside the multiple effects that shape the given. What such a move does, as Teresa Ebert has argued, is maintain the "matterism" of the eighteenth century that equates the material with "thingness" and makes experience the limit of knowing.[109] Whether this discursive immanentism it is put forward in a Heideggerian language as in Sara Ahmed's essay which argues that "standpoint" determines what "matters,"[110] or a Derridean one such as in Cheah's essay in which the material is represented as "the inappropriable other,"[111] or as in Elizabeth Grosz's use of Agamben's language of "bare life" as an affirmative "materiality that life and the nonliving share,"[112] makes no difference. *New Materialisms* returns to the ahistorical eighteenth century "matterism" as a "new" materialism to understand the contemporary crisis of capitalism and its knowledge industries. The claim that this is "new" amounts to the mantra that if matter is no longer considered "passive and inert" but "productive and resilient,"[113] as in the vitalism of the Spinoza-Nietzsche-Bergson-Deleuze axis, this will "reconfigure our models of society and the political"[114] in "new" ways that are not "determinable" in terms of "cause and effect."[115] What this claims is that the material is "aleatory" and "immaterial"[116] as it can only be known at the level of its effects. The question is: Is matter, as in the tiresome pun, just what "matters" to people (belief) which changes with cultural changes in values, or, do discourses change with changes in the mode of production of material life? Why is the fundamental assumption in *New Materialisms* that materialism—in the sense of positive and reliable knowledge "outside" ideology which explains the cultural in terms of class—can no longer be considered explanatory? The "why" here is of course the matter of class.

The argument of the new materialists for turning to the material so as to construct a more ethical theory that promises to change the cultural real turns materialism on its head, to paraphrase Marx on Hegel. Materialism, as Marx explains, always consists of looking "more closely" at how "social being determines [. . .] consciousness"[117] rather than the other way around, especially at such times when men and women "seem engaged in revolutionising themselves and things, in creating something entirely new,"[118] while forgetting that history is only made "under circumstances directly found, given and transmitted from the past."[119] By making the effectivity of materialism consist of an ethical break with culturalism, Coole and Frost construct a culturally immanent history of ideas that makes history seem like a series of contingent "events" that elude explanation in terms of class. And yet, the reason why materialism has suddenly become the "new"—if that is to mean more than a marketing gimmick to repackage the same old culturalism in crisis—needs to be explained in relation to class and the new class struggles of transnational capitalism in crisis.

BIBLIOGRAPHY

Ahmed, Sara. "Orientations Matter." In Coole and Frost, *New Materialisms*, 234–57.

Bennett, Jane. *Vibrant Matter: A Political Ecology of Things*. Durham, NC: Duke University Press, 2010.

Bolt, Barbara. "Material Thinking and the Agency of Matter." *Studies in Material Thinking* 1, no. 1 (April 2007): 1–4. Accessed January 1, 2016. https://www.materialthinking.org/papers/37.

Braidotti, Rosi. "The Politics of 'Life Itself' and New Ways of Dying." In Coole and Frost, *New Materialisms*, 201–18.

Cheah, Pheng. "Non-Dialectical Materialism." In Coole and Frost, *New Materialisms*, 70–91.

Connolly, William E. "Materialities of Experience." In Coole and Frost, *New Materialisms*, 178–200.

Coole, Diana and Samantha Frost. "Introducing the New Materialisms." In Coole and Frost, *New Materialisms*, 1–43.

Coole, Diana and Samantha Frost, eds. *New Materialisms: Ontology, Agency, and Politics*. Durham, NC and London: Duke University Press, 2010.

De Fontenay, Elisabeth. *Without Offending Humans: A Critique of Animal Rights*. Minneapolis: University of Minnesota Press, 2012.

Deleuze, Gilles and Félix Guattari. *A Thousand Plateaus: Capitalism and Schizophrenia*. Minneapolis: Minnesota University Press, 1987.

Deleuze, Gilles. *Pure Immanence: Essays on A Life*. New York: Zone Books, 2005.

De Man, Paul. *The Resistance to Theory*. Minneapolis: Minnesota University Press, 1986.

Derrida, Jacques. *Politics of Friendship*. London and New York: Verso, 1997.

---. *Specters of Marx*. New York: Routledge, 1994.

---. "Typewriter Ribbon: Limited Ink (2) ('within such limits')." In *Material Events: Paul de Man and the Afterlife of Theory*, edited by Tom Cohen, Barbara Cohen, J. Hillis Miller, and Andrzej Warminski, 277–360. Minneapolis: Minnesota University Press, 2001.

Dolphijn, Rick and Iris van der Tuin. *New Materialism: Interviews and Cartographies*. Ann Arbor, MI: Open Humanities Press, 2012.

Ebert, Teresa L. *Ludic Feminism and After: Postmodernism, Desire, and Labor in Late Capitalism*. Ann Arbor: Michigan University Press, 1996.

Edwards, Jason. "The Materialism of Historical Materialism." In Coole and Frost, *New Materialisms*, 281–98.

Grosz, Elizabeth. "Feminism, Materialism, and Freedom." In Coole and Frost, *New Materialisms*, 139–57.

Haraway, Donna. *When Species Meet*. Minneapolis: Minnesota University Press, 2008.

Iovino, Serenella and Serpil Oppermann. "Material Ecocriticism: Materiality, Agency, and Models of Narrativity." *Ecozon@* 3, no. 1 (2012): 75–91.

Jameson, Fredric. *The Political Unconscious: Narrative as a Socially Symbolic Act*. Ithaca, NY: Cornell University Press, 1988.

Khan, Gulshan Ara. "Vital Materiality and Non-Human Agency: An Interview with Jane Bennett." In *Dialogues with Contemporary Political Theorists*, edited by Gary Browning, Raia Prokhovnik, and Maria Dimova-Cookson, 42–57. New York: Palgrave Macmillan, 2012.

Lenin, V. I. "On the Significance of Militant Materialism." *On Culture and Cultural Revolution*. Moscow: Progress Publishers, 1970.

Marx, Karl and Frederick Engels. *The German Ideology*. Moscow: Progress Publishers, 1976.

---. *The Marx-Engels Reader*. Edited by Robert C. Tucker. New York and London: W. W. Norton and Co., 1978.

Marx, Karl. *Capital: A Critique of Political Economy*. Vol. 1. Translated by Ben Fowkes. London: Penguin Books, 1976.

---. "Contribution to the Critique of Hegel's Philosophy of Law," *Collected Works of Karl Marx and Friedrich Engels*. Vol. 3, 175–87. Moscow: Progress Publishers, 1975.

---. *Economic and Philosophic Manuscripts of 1844*. Moscow: Progress Publishers, 1974.

---. *The Eighteenth Brumaire of Louis Bonaparte*. In Marx and Engels, *The Marx-Engels Reader*, 594–617.

---. *Grundrisse: Foundations of the Critique of Political Economy*. Translated by Martin Nicolaus. New York: Penguin, 1993.

---. "Preface to *A Contribution to the Critique of Political Economy*." In Marx and Engels, *The Marx-Engels Reader*, 3-6.

---. "Theses on Feuerbach." In Marx and Engels, *The German Ideology*, 615–17.

Mies, Maria. *Patriarchy and Accumulation on a World Scale: Women and the International Division of Labour*. London: Zed Books, 1986.

Roelvink, Gerta. "Rethinking Species-Being in the Anthropocene." *Rethinking Marxism* 25, no. 1 (2013): 52–69.

NOTES

1. Paul De Man, *The Resistance to Theory* (Minneapolis: Minnesota University Press, 1986).
2. Jacques Derrida, *Politics of Friendship* (London and New York: Verso, 1997).
3. Gilles Deleuze, *Pure Immanence: Essays on A Life* (New York: Zone Books, 2005).
4. Michael Hardt and Antonio Negri, *Multitude: War and Democracy in the Age of Empire* (New York: Penguin Press, 2004).
5. Slavoj Žižek, *The Sublime Object of Ideology* (London and New York: Verso, 1989).
6. Donna Haraway, *When Species Meet* (Minneapolis: University of Minnesota Press, 2008).
7. Rick Dolphijn and Iris van der Tuin, *New Materialism: Interviews and Cartographies* (Ann Arbor: Open Humanities Press, 2012).
8. Jane Bennett, *Vibrant Matter: A Political Ecology of Things* (Durham, NC: Duke University Press, 2010).
9. Barbara Bolt, "Material Thinking and the Agency of Matter," *Studies in Material Thinking* 1, no. 1 (April 2007): 1–4, accessed January 1, 2016, https://www.materialthinking.org/papers/37; and Serenella Iovino and Serpil Oppermann, "Material Ecocriticism: Materiality, Agency, and Models of Narrativity," *Ecozon@* 3, no. 1 (2012): 75–91.
10. Jacques Derrida, "Typewriter Ribbon: Limited Ink (2) ('within such limits')," in *Material Events: Paul de Man and the Afterlife of Theory*, ed. Tom Cohen, Barbara Cohen, J. Hillis Miller, and Andrzej Warminski (Minneapolis: Minnesota University Press, 2001), 281.
11. On posthumanism see my, "Animal Matters, Sublime Pets, and Other Posthumanist Stories," *The Red Critique* 14 (Winter/Spring 2012), accessed January 1, 2016, http://redcritique.org/WinterSpring2012/animalmatterssublimepetsandotherposthumaniststories.htm.
12. Gerta Roelvink, "Rethinking Species-Being in the Anthropocene," *Rethinking Marxism* 25, no. 1 (2013): 52–69.
13. Teresa L. Ebert, *Ludic Feminism and After: Postmodernism, Desire, and Labor in Late Capitalism* (Ann Arbor: Michigan University Press, 1996), 24–36.
14. Fredric Jameson, *The Political Unconscious: Narrative as a Socially Symbolic Act* (Ithaca, NY: Cornell University Press, 1988), 45.
15. Pheng Cheah, "Non-Dialectical Materialism," in *New Materialisms: Ontology, Agency, and Politics*, ed. Diana Coole and Samantha Frost (Durham, NC and London: Duke University Press, 2010), 70– 91.
16. Ibid., 73.
17. Ibid., 72.
18. Ibid., 71.
19. Ibid., 71.
20. Ibid., 71.
21. Ibid., 79.
22. Ibid., 71.
23. Ibid., 80.
24. Ibid., 81.
25. Karl Marx, "Contribution to the Critique of Hegel's Philosophy of Law," *Collected Works of Karl Marx and Friedrich Engels*, vol. 3, 1843–44 (Moscow: Progress Publishers, 1975), 182.
26. Karl Marx, "Theses on Feuerbach," in Karl Marx and Frederick Engels, *The German Ideology* (Moscow: Progress Publishers, 1976), 616.
27. V. I. Lenin, "On the Significance of Militant Materialism," in *Culture and Cultural Revolution* (Moscow: Progress Publishers, 1970), 184–97. Lenin argues that "militant materialism" is above all necessary to counter the way that "the sharp upheavals which modern natural science is undergoing very often gives rise to reactionary philosophical trends" (192) that make "democratic Left or ideologically socialist publicists"

(187) as well as "modern scientific critics of religion [. . .] ideological slaves of the bourgeoisie" (189) by "openly helping the exploiters to replace the old, decayed religious superstitions by new, more odious and vile superstitions" (190) such as the "unreasonable worship" of "so-called modern democracy" (191), which is "nothing but the freedom to preach whatever is to the advantage of the bourgeoisie" (191).

28. Jacques Derrida, *Specters of Marx* (New York: Routledge, 1994), 168–69.

29. Diana Coole and Samantha Frost, "Introducing the New Materialisms," in Coole and Frost, *New Materialisms*, 3.

30. Ibid., 6.

31. Cheah, "Non-Dialectical Materialism," 73.

32. Ibid., 73.

33. Ibid., 72.

34. Ibid., 73.

35. Ibid., 73.

36. Ibid., 89.

37. Ibid., 73.

38. Ibid., 78.

39. Ibid., 80.

40. Ibid., 80.

41. Ibid., 81.

42. Ibid., 80, citing Jacques Derrida, *Politics of Friendship* (London and New York: Verso, 1997), 68–69.

43. Gulshan Ara Khan, "Vital Materiality and Non-Human Agency: An Interview with Jane Bennett," in *Dialogues with Contemporary Political Theorists* , ed. Gary Browning, Raia Prokhovnik, and Maria Dimova-Cookson (New York: Palgrave Macmillan, 2012), 46.

44. Cheah, "Non-Dialectical Materialism," 89.

45. Ibid., 89.

46. Karl Marx, *Grundrisse: Foundations of the Critique of Political Economy* , trans. Martin Nicolaus (New York: Penguin, 1993), 100.

47. Ibid., 101.

48. Ibid., 100.

49. Cheah, "Non-Dialectical Materialism," 87

50. Ibid., 87.

51. Derrida, "Typewriter Ribbon," 278.

52. Ibid., 281.

53. Ibid., 278.

54. Ibid., 277.

55. Ibid., 292.

56. Cheah, "Non-Dialectical Materialism," 77.

57. Ibid., 89.

58. Ibid., 80.

59. Ibid., 81.

60. Ibid., 71.

61. Ibid., 71.

62. Elisabeth De Fontenay, *Without Offending Humans: A Critique of Animal Rights* (Minneapolis: University of Minnesota Press, 2012), 48–49.

63. Marx, "Theses on Feuerbach," 616.

64. "[T]he world-historical co-operation of individuals [. . .] will be transformed by this communist revolution into the control and conscious mastery of these powers, which, born of the action of men on one another, have till now overawed and governed men as powers completely alien to them. Now this view can be expressed again in speculative-idealistic, i.e. fantastic, terms as 'self-generation of the species' ('society as the subject'), and thereby the consecutive series of interrelated individuals connected with each other can be conceived as a single individual, which accomplishes

the mystery of generating itself. It is clear here that individuals certainly make one another, physically and mentally, but do not make themselves," Ibid., 59.

65. Ibid., 616.
66. Marx and Engels, *The German Ideology*, 37.
67. Karl Marx, *Capital: A Critique of Political Economy*, vol. 1, trans. Ben Fowkes (London: Penguin Books, 1976), 164.
68. Karl Marx, *Economic and Philosophic Manuscripts of 1844* (Moscow: Progress Publishers, 1974), 67.
69. Ibid., 68.
70. Coole and Frost, "Introducing the New Materialisms," 37.
71. Ibid., 3.
72. Jameson, *The Political Unconscious*, 45.
73. Coole and Frost, "Introducing the New Materialisms," 3.
74. Ibid., 3.
75. Ibid., 6.
76. Ibid., 5.
77. Rosi Braidotti, "The Politics of 'Life Itself' and New Ways of Dying," in Coole and Frost, *New Materialisms*, 206.
78. William E. Connolly, "Materialities of Experience," in Coole and Frost, *New Materialisms*, 191.
79. Ibid., 194.
80. Braidotti, "The Politics of 'Life Itself'," 211.
81. Maria Mies, *Patriarchy and Accumulation on a World Scale: Women and the International Division of Labour* (London: Zed Books, 1986), 37.
82. Connolly, "Materialities of Experience," 195.
83. Cheah, "Non-Dialectical Materialism," 74.
84. Braidotti, "The Politics of 'Life Itself'," 211.
85. Ibid., 206.
86. Ibid., 213.
87. Ibid., 213.
88. Jason Edwards, "The Materialism of Historical Materialism," in Coole and Frost, *New Materialisms*, 297.
89. Ibid., 282.
90. Ibid., 282.
91. Ibid., 283.
92. Ibid., 295.
93. Ibid., 295.
94. V. I. Lenin, "On the Significance of Militant Materialism," in *Culture and Cultural Revolution* (Moscow: Progress Publishers, 1970), 184–97.
95. Coole and Frost, "Introducing the New Materialisms," 3.
96. Ibid., 3.
97. Ibid., 13.
98. Ibid., 3.
99. Cheah, "Non-Dialectical Materialism," 72.
100. Ibid., 73.
101. Ibid., 72.
102. Ibid., 71.
103. Coole and Frost, "Introducing New Materialism," 3.
104. Cheah, "Non-Dialectical Materialism," 73.
105. Ibid., 81.
106. Ibid., 73.
107. Karl Marx, "Hegel's Philosophy of Law," 182.
108. Cheah, "Non-Dialectical Materialism," 75.
109. Ebert, *Ludic Feminism* , 24–36.
110. Sara Ahmed, "Orientations Matter," in Coole and Frost, *New Materialisms*, 234–57.

111. Cheah, "Non-Dialectical Materialism," in *New Materialisms*, 80.
112. Elizabeth Grosz, "Feminism, Materialism, and Freedom," in Coole and Frost, *New Materialisms*, 142.
113. Coole and Frost, "Introducing New Materialism," 7.
114. Ibid., 13.
115. Ibid., 14.
116. Ibid., 14.
117. Karl Marx, "Preface to *A Contribution to the Critique of Political Economy*," in Karl Marx and Frederick Engels, *The Marx-Engels Reader*, ed. Robert C. Tucker (New York and London: W. W. Norton and Co, 1978), 4.
118. Karl Marx, *The Eighteenth Brumaire of Louis Bonaparte*, in Marx and Engels, *The Marx-Engels Reader*, 595.
119. Ibid., 595.

FOUR

Mind over Matter and Other Posthumanist Feminist Tales

Julie Torrant

Recently, there has been a turn away from textualist and culturalist theory in feminism and the emergence of "new materialist" feminisms. Represented by the work of Elizabeth Grosz, Rosi Braidotti, and others, this turn in theory has come in response to the deepening contradictions and crises of capitalism that are having profound effects on women worldwide—material contradictions that constitute problems outside the text that are not resolvable by a change in cultural values. Contemporary capitalism is shaped by a systemic crisis of "overproduction": the tendency for constant capital to increase relative to variable capital in the production process thus causing the rate of profit to fall because further investment does not serve to realize greater "returns" (a crisis of profitability). The increase in the ratio of constant to variable capital is driven by the inevitable pressures of competition between individual capitalists, who introduce labor-saving means of production in order to increase the ratio of surplus value (profit) to necessary wages within the working day and thus enable them to price their goods more cheaply while still reaping "surplus" profits in relation to their competitors; however, once their competitors introduce these new means of production, and constant capital (investment tied up in plant and equipment) is increased in relation to variable capital (wages), it makes it harder for all capitalists to profit because it is labor that is the source of value. In other words, as Marx writes, "Thus the capitalist will have gained nothing more by his efforts than the obligation to furnish a greater product in the same labour-time; in a word, more difficult conditions for the profitable employment of his capital."[1] At the same time, this dynamic means an increase in the rate of exploitation of workers since the proportion of surplus to necessary labor means that the labor they perform for wages and thus to meet their needs is proportionally less than the labor they perform that is appropriated by capitalists in the form of surplus labor (the source of profits). As capitalism increases social inequality through this mechanism of creating more surplus-value and more workers find themselves unable to meet their needs, they are driven into social struggles that demand resources at a time when the rate of profit is declining and capital has less need for labor (within the nation). Capitalism, put another way, produces an "inverted" world in which all human endeavors turn into their opposite despite the will of the doers. Capitalists aim to increase their rate of profit, but as a class produce a crisis of profitability, and workers attempt to work more efficiently with new means of production in order to meet their needs, but within the current mode of production and its private property relations their labor facilitates the development of ever newer forms of property which are estranged from them and thus actually deepens the historically produced conditions of scarcity which in turn drives them to accept deepening exploitation.

Amongst the forms contemporary capital's class struggles have taken is a decades-long, multi-faceted "neo-liberal" campaign to increase the rate of exploitation of workers and working families and to shift the burden of reproduction of the workforce to the working class and thus defer the increasingly deep, global economic crisis of capitalism. This crisis and the crisis-management policies it has given rise to have produced intransigent "recessions" and "depressions" as well as financial "crises" (followed by bailouts that further burden workers with regressive tax structures and shrinking government services or "austerity") such as those that have plagued Greece, that have made global capitalism increasingly volatile and violent for workers across the globe. Global workers are subject, if differentially, to wars underpinned by historically produced scarcity of and struggles over resources, un- and underemployment, austerity, ideologically fueled wars on the poor and "others" within nations as well as economic and political refugees across national borders, and the depredations of uneven, imperialist development which cannot account for the necessity for safe working conditions, clean air and water, healthy food, health care, and the other basic and historically produced needs of twenty-first century workers. The flip side of neo-liberal austerity is the turn to "finance" and the production of financial speculation and bubbles (because of the "overproduction" of capital—capital that cannot be profitably invested in production) that at best temporarily cover over but ultimately deepen the systemic crisis.

These developments in production and the economy have meant that women globally have been increasingly drawn directly into the structures of capitalism and as such have been subject to uneven and combined development produced by class-divided social relations of production. Within these social relations a tiny minority of ruling class women have joined the ruling (capitalist) class and reap the benefits of control over the surplus labor of the vast majority of women who have been drawn, *en masse*, into the capitalist workforce, but under conditions where they are subject to deepening inequalities across the class divide as well as across the social fissures of gender, race, nationality, sexuality, age, and (dis)ability and the divisions of labor that underpin these ideological differences.

The "new materialist" feminisms are, broadly speaking, responses to real problems that have been produced by contemporary capitalism and its conflictual drive for profits in the context of the development of new forms of property (such as "hedge funds" and other financial capital)—forms of property which can make it appear that capital develops independently of the daily lives and struggles of workers and their labor—and a deepening social as well as economic crisis. However, through their re-conceptualization of what is material as "the forces, processes, capacities, and resiliencies with which bodies, organisms, and material objects act both independently and in response to discursive provocations and constraints," in other words, by positing the material as the autonomy of "matter"—and thus as a "supplement [to] cultural and discursive analysis of social and political phenomenon" rather than a re-thinking of the erasure of class and labor from cultural and social theory—these "new materialist" feminisms are ultimately part of the class struggles on behalf of capitalists which justify the current material (property) relations and aid the class-interested containment of collective struggles on the part of workers.[2] Contrary to their self-representations, "new materialist" feminisms are disenabling forms of spiritualism that displace explanatory critique of the emergent material conditions with strategies of enchanted affective adaptation and survival. These spiritualisms demonize, marginalize and suppress critique and, as such, work to dismantle materialist feminism's primary conceptual tool for social transformation in favor of an updated ruling-class ideology that prioritizes crisis-management over social transformation. To avoid merely reproducing sophisticated forms of the survivalism and "prepperism" that have emerged as individualistic coping responses to economic crisis and austerity, I argue that feminism needs to return to historical materialism in the tradition of Marx, Engels, and Kollontai to understand social life in terms of its root relations and aid in the struggles to bring about social transformation.

Exemplary of the "new materialist" feminism is Rosi Braidotti's writing on "the politics of 'life itself,'" a theory which she organizes around the trope of "sustainability." Sustainability, a concept in ecology for living within natural limits, becomes in these writings a means of reconceiving the historical social relations of capitalism as if they were the unchangeable, underlying existential limit-situation of "life itself." The politics of "life itself" and the "new materialist" focus on seeking a sustainable feminism within this new, more "realist" approach to material reality, is a form of feminist theory and politics which is ultimately the already familiar theory and politics of reparative reading. Why is this significant? As Ellis Hanson suggests in a review of Sedgwick, "Faced with the depressing realization that people are fragile and the world hostile, a reparative reading focuses not on the exposure of political outrages that we already know but rather on the process of reconstructing a sustainable life in their wake."[3] In other words, reparative analysis begins not with critique of the so-called already known and presumably known to be unchangeable, but by focusing on how to live within the already-known-to-be hostile world. Such a theory of the social begins and ends by reducing knowledge to a matter of how to cope, how to feel, how to exist, etc. within what is taken to be unchangeable. The effect of this focus on "sustainability" within hostility is that social transformation—which requires the production of knowledge of what needs to be transformed—is treated as impossible. Abandoning the project of transformation, I argue, is a sign of the way dominant "materialist" feminism—under the guise of "new materialism"—has increasingly abandoned the project of women's emancipation from exploitation, and, in the interests of capital, instead translates austerity measures into a theoretical discourse of getting by on less.

At the core of Braidotti's theory of "sustainable feminism" and "life politics" is a "new materialist" understanding of "life." For Braidotti, life is made up of two parts—*zoē* and *bios*. Zoē, "life as absolute vitality," is the spiritual and bios is the "bio-organic" body which sets limits on the spiritual life force.[4] Braidotti writes, "Zoē, or life as absolute vitality, however, is not above negativity, and it can hurt. It is always too much for the specific slab of enfleshed existence that single subjects actualize. It is a constant challenge for us to rise to the occasion, to catch the wave of life's intensities and ride it."[5] Thus for Braidotti, the source of social contradictions is the conflict between *zoē*, that is, absolute vitality or spiritual life force, and our bio-organic bodies. As a result, Braidotti's new materialism bypasses the ensemble of social relations and historical conditions that produce social contradictions in capitalism and presents contradictions as transhistorical and existential conditions of life as such. On this logic, our absolute vitality comes into the world and reaches the limit of the body and this causes us "pain." According to this narrative, however, there is no real way to compensate for pain. This explanation of pain is an example of bypassing the social. As such it is an accommodationist block to changing the conditions that produce suffering.

Thus, according to the underlying mysticism of "new materialism," when political-economic refugees from the Middle East and northern Africa drown, suffocate, or otherwise perish in attempts to cross into Europe, it would seem that their spiritual life force, which drove them to seek a better life for themselves and their families, came up against the limits of their bio-organic bodies and this is just one of the innumerable examples of the inevitable "pain" experienced by individuals who come up against such a conflict between *zoē* and *bios*. Moreover, on this view, citizens of Europe, the United States, and beyond are encouraged to view their own "pain" upon witnessing and learning about these deaths as a matter of an "intensity" that they should "ride" (feel) as much as possible in order to enable themselves to "deal with" this tragic situation and help to create "sustainable" change, which implies reforms to immigration policy. What this approach does not address or enable, however, is why and how to understand the class relations that have produced such "migrants" and why and how these class relations also shape the ideological conditions that produce people as completely expendable surplus labor or as, in essence, trash to be disposed of as efficiently as possible with the least disruption and cost as possible. This is the way in which "reparative reading" does the necessary ideological work for capital: it focuses on emotional response to the effects of exploitation but obscures underlying causes in the structural relations of capitalism, or the need to change them, and teaches workers how to adjust to and work within the conditions of their own exploitation.

In fact, as with all the popular articulations of "materialism" today, Braidotti's theory is not actually an extension of materialism, but a break from it. The "new materialism" is really idealism. Materialism and idealism, as Lenin clarifies, split on the question of "Whether nature, matter, the physical, the external world be taken as primary, and mind, spirit, sensation (experience—as the widespread terminology of our time has it), the psychical, etc., be regarded as secondary."[6] Materialism is the recognition of the external world as primary and determinative of mind and experience, whereas idealism posits the reverse. Historical materialism, or determination by the mode of production, is the explanation of sense experience as the outcome of production. As humans produce their means of life by working upon nature within a particular mode of production they also produce themselves. Thus, not only are the objects we see, hear, feel, taste, touch, etc., the product of the labors of workers today *and* of the history of labor these labors build on, but the senses, our ways of perceiving, are also themselves shaped by production.

For instance, the food that we eat is what it is because of the way agricultural laborers work on nature through the mediation of the existing means of production. The foods we find in the supermarket today are the product of industrialized agricultural production using large, computerized machinery, substances produced in food science labs, genetically modified seeds, chemical fertilizers, and so forth. Moreover, the way we perceive these foods is shaped by the capitalist mode of production and its (private) property relations. Consumers are enticed to buy and eat very unhealthy food full of saturated fats and large amounts of salt and sugar that have been designed to trigger their cravings for sweet, fatty, and salty foods combined and these foods are "seen" as "treats" that meet (unmet) needs.[7] At the same time, many consumers are becoming wary of the food they see in the supermarket because they are increasingly aware of the way the production of these foods is not determined by their need for nutritious as well as inexpensive and satisfying food, but by profit. Thus, we see the emergence of a "slow food" movement. However, it is impossible for workers to simply reject highly processed foods, not simply for economic reasons, but also because of the de-limitation of our sense experience of food and, in essence, the social psychology of eating, which at present involves a complex relation wherein food serves as a compensation for unmet needs such as needs for satisfying work, health, and intimacy that are produced within capitalism. As Marx writes in *The Economic and Philosophical Manuscripts,* "The abolition of private property is therefore the complete *emancipation* of all human sense and qualities, but it is this emancipation precisely because these senses and attributes have become, subjectively and objectively, *human.*"[8] In other words, the way we perceive under capitalist private property relations is produced and thus de-limited by these relations such that it reflects the narrow interests of capital in profit rather than representing workers' collective potential for all-sided experience and enjoyment.

In contrast to this way of explaining sense experience, in "new materialism" experience is turned on its head because on this view "life" exceeds our ability to "know" it. Our experience, for "new materialists," is not the reflection of the objective world, but is a cultural construct, a matter of what we "believe" or "feel." But what they call "materialism" (the reality that transcends our experience) is really objective idealism—the view that there is a transcendental force—whether "Spirit" or "God" or "matter/life" that determines our experience and cannot be grasped by (human) consciousness or "culture." Our culture, as historical materialism argues, is the repository of ideas through which we make sense of our experience and "fight out" the class struggles that are rooted in the economic base of labor relations. The "new materialists" are opposed to the idea that culture is a superstructural reflection of the economic base (of labor relations). And yet, their concept of the material (awesome, mindless, etc.) is in fact a mediated reflection of the new property forms of transnational capitalism in which capital seems to take on a life of its own untethered from the daily life in which workers are forced to accommodate themselves to its drive. Hence this form of "new materialism" gives rise to the notion of a natural state of "precarity" and valorizes practices of "sustainability" such as "sustainable" food in the absence of an explanatory critique of the production of the experience/s of food. As Marx argues, from the perspective of historical materialism, "[w]e see how the history of industry and the established objective existence of industry are the open book of man's essential powers, the perceptibly existing human psychology."[9] In contrast to historical materialism, new materialism advocates a spiritualism that displaces social psychology with "[a] *psychology* for which this book [the history of industry and the established objective existence of industry], the part of history existing in the most perceptible and accessible form, remains a closed book."[10] In other words, "new materialism" reifies sense experience from the conditions that produce and shape it.

Materialism is not the experience that exceeds conceptuality—a Kantian theory of the material that has come to dominate cultural theory, especially as it conceives of "life." This notion of materialism merely reifies sense experience from its social basis, it cannot explain it. In spite of claims to the contrary, Braidotti is Kantian about the material because she sees it as a sublime excess. Life, Braidotti writes,

> is experienced as inhuman because it is all too human, obscene because it lives on mindlessly. Are we not baffled by this scandal, this wonder, this zoē, that is to say, by an idea of life that exuberantly exceeds bios and supremely ignores logos? Are we not in awe of this piece of flesh called our 'body,' of this aching meat called our 'self' expressing the abject and simultaneously divine potency of life?[11]

According to Braidotti, what exceeds the individual body is *zoē*—the spiritual life force, which we should not understand conceptually (by seeking to explain the conditions that shape it) but which we should worship. This is a sentimental call in the name of anti-instrumentalism for the re-enchantment of life that obscures the way the individual is determined not by what Braidotti calls "divine potency" but by the social relations of production. And, like all "anti-instrumental" arguments, Braidotti's ends up affirming a species of the sublime: a mode of affective non-knowing that resists rationality and thereby blocks the formation of the workers into a class-for-itself—the collective agency of social change—on behalf of capital.

Thus, having rejected the necessity of being able to conceptualize effects in relation to their causes, Braidotti proceeds to declare that the effects of living in the ruins of capitalism—especially disasters like 9/11—defy all reason and are impossible to understand, and she concludes that what is now necessary is not collective praxis to address the social relations which condition the unequal situations of tragedy, but an individual ethics of affirmation. She writes:

> This is the road to an ethics of affirmation, which respects the pain but suspends the quest for both claims and compensation. The displacement of the "zoē"-indexed reaction reveals the fundamental meaninglessness of the hurt, the injustice, or injury one has suffered. "Why me?" is the refrain most commonly heard in situations of extreme distress. The answer is plain: actually, for no reason at all. Why did some go to work in the World Trade Center on 9/11 while others missed the

> train? Reason has nothing to do with it. That's precisely the point. We need to delink pain from the quest for meaning.[12]

Following her predictable rejection of concepts and reason, in the guise of a sermon on "selflessness" and acceptance of "what is," Braidotti here once again rejects the abstract in favor of the errant concrete and takes as a presupposition the individual. For it is of course from the starting point of individuals and their loss that we cannot understand and explain such historical events as 9/11. From the perspective of the individual, such events are indeed random and inexplicable, but from a historical perspective they are determined. It is deep global inequities that provided the conditions of possibility for the 9/11 attacks. To celebrate the individual perspective and the inability to grasp historical necessity based on that individual perspective is not only to celebrate ignorance, but to naturalize the limits of workers and how they are thrust into the position of individuals who must compete on the market for work while leaving it the prerogative of the owners to organize the totality to the benefit of a few at the expense of the many.

Central to Braidotti's enchanted materialism is her claim that affectivity "is what activates an embodied subject, empowering him or her to interact with others."[13] However, she writes, "a subject can think/understand/do/become no more than what he or she can take or sustain within his or her embodied, spatiotemporal coordinates."[14] Thus, the ethical subject is the one who learns to endure his or her maximum zoē/bios intensity because such endurance leads to "sustainable transformations," the degree of change an individual can bear.[15]

But the consequences of affirmative ethics are deeply problematic when considered in relation to the material conditions of working class families who have been subject to capital's attempts to manage its crisis of overproduction and the proliferating crises this attempt has produced. As I have begun to discuss, the overproduction of capital is the result of the competition between and among capitalists, who introduce labor-saving technologies in order to win the competition for market share and profits. Labor-saving technologies constitute for workers, however, an increase in the ratio of surplus to necessary labor, or an increase in the rate of exploitation, even though they do not necessarily mean a drop in nominal wages. Empirical evidence shows how deep the exploitation of families has become in recent decades. For instance, according to *State of Working America*, wages over the last three decades have either stagnated or have grown very little even as worker productivity has sharply increased. Specifically, the report finds that "Worker hourly wages at the 10th percentile were essentially flat, while median wages grew about 0.3 percent each year in this 28-year period" of 1979–2007.[16] In the wake of the more recent 2007 crisis, worker productivity has sharply increased,[17] while wages have fallen. The increase in the rate of exploitation means a deepening of the class divide as the increasing productivity of workers produces a greater mass of surplus value for the owners while depriving the workers of access to the wealth and resources they collectively produce. In other words, the contradiction between what is and what could be widens.

Moreover, labor saving technologies also function to make "redundant" large numbers of workers. In the "advanced" capitalist countries over the last several decades, this has meant the loss of large numbers of well-paying manufacturing or other jobs, which has also meant the end (more or less) of the family wage for the vast majority of workers. As a result, although the nominal (or even, for some, real) wages of working families has risen with the increase in hours of the family working day (i.e., the productive and wage labor of all family members, not their reproductive labor), given that it takes the wages and thus labor time of two or more workers to support the family, the exploitation of these families has deepened. This is because they are producing a greater proportion of surplus labor (profits) in relation to necessary labor (wages) than the comparable single-wage earner family did when fewer women worked for wages. As Mishel et al. observe,

> Because American households added so many more hours to the paid labor force between 1979 and 2007 and because the later 1990s provided a welcome period of strong across-the-board wage growth, the full extent of the wage disaster for the majority of American workers for *most* of the years between 1979 and 2007 has often been underappreciated. Between 1979 and 1995, for example, hourly wages at the 10th percentile and the median fell at average annual rates of 0.9 percent and 0.2 percent . . . (Undoubtedly, Americans started working more paid hours beginning in 1979 in part as a coping strategy to ensure some income growth despite poor wage performance).[18]

For the upper strata of working families this has meant an increase in income and resources and thus a higher standard of living, and for this stratum of women more opportunities for fulfilling work and more equality of opportunity in relation to men in the workplace and beyond. However, as Elizabeth Warren and Amelia Warren Tyagi argue in *The Two-Income Trap: Why Middle-Class Parents are Going Broke*, this new situation brings with it risks and "vulnerability" for even the upper strata of working families (what they call the "middle-class") because when there were (among primarily white, "middle-class" strata families) single-earner families, if the family sustained some kind of economic blow (say the loss of a job due to layoff or illness or a partial loss of income based on a pay-cut), then the second adult in the family (usually the woman/wife) could enter the workforce.[19] In other words, the "reserve" worker, who, from the perspective of capital formed part of the reserve army of labor, could serve as a buffer against various contingencies that might mean a loss of income for the family. However, once most families have two wage-earners, and, importantly, as Warren and Tyagi point out, the second income is *not* being used for "luxuries" or "extras" such as vacations, but rather is needed to pay bills such as the mortgage, health insurance, and tuition,[20] then in case of the loss of a job or income, for whatever reason, the family has no "reserve" worker, and the family loses what Warren and Tyagi call "the most important part of [the family's] safety net."[21] Thus, such families are, inevitably, whether they realize it or not, in a permanently "precarious" state. They are, in Warren and Tyagi's words, "vulnerable" to financial crisis.[22] In this context, the idea that "precarity" is the natural state of things, or, in Braidotti's terms, there is a perpetual state of spiritual "excess" or "intensity" that one must learn to "ride" with, works to naturalize this precarious financial state and mode of living and working under intense exploitation that even the more privileged women and men, whom Braidotti and other "new materialists" reach in their writings, are subject to in contemporary capitalism.

To give another example, partners who work full time for wages and then are caring for children and aging parents have such limited "free time" as well as psychological stresses and strains that it makes having a fulfilling relationship impossible—rather than a matter of an "intensity" to be ethically mastered. Alongside of the trends towards deepening exploitation and longer working hours, rates of violence against women have increased dramatically[23] and suicide is now the tenth leading cause of death in the US.[24] That the spouses in working class families are increasingly emotionally strained and often alienated from one another is not a transhistorical effect of their embodied state as it confronts a "divine" life force in *zoē*, but an effect of their deepening exploitation. To posit their connection and disconnection with *bios-zoē* is to de-historicize their pain and alienation as individuals and as a couple. It is to cut off affect from its social conditions and then insist on its affirmation.

"New materialist" writings ideologically "invert" the sense experience of financial insecurity and other forms of contradiction as pleasure and a positive ethical stance rather than enabling their readers to grasp why this state exists and how it is related to the underlying profit-driven class relations and material conditions produced by these class relations. The "intensity" of living on the edge of one's limits (financial, psychological, physical, etc.) is not an ethical necessity for a more just world, it is a ruling class ideology that makes sense of the insecurity of living within a state of historically produced scarcity as the pleasure and positive social benefits of living in existential flux. This ideology serves capital's need to manage its crisis and keep workers from rejecting the unnecessary suffering such scarcity produces for all workers (if differentially). Put another way, living in contemporary capitalism can certainly be "intense"; however, it is in the interests of working people and working families not to merely learn to live with this intensity now that it is increasingly evident to all workers that they are living in a crisis-prone state even if/when they themselves are not in immediate crisis, but to understand why and how it is necessary and possible to establish a society and social relations that provide the basis for more secure and safe living and working conditions not only for themselves, but for their fellow workers at home and beyond. What this means for women is that the gains that some women have made in being drawn into the labor force are, at best, established on a kind of economic "fault line" of economic crisis and thus contrary to ruling class ideologies, women's interests are not with capital, national or global, but with other workers along with the people who are made "redundant" in contemporary capitalism.

Working more hours is a matter of "making do," not existential intensity, and it is this making do under conditions of deepening exploitation that all working class families—gay and straight, white and of color, native and international—have been forced to do and which affects women profoundly. As Marx and Engels explain in their analysis of the global development of capitalism,

> The less the skill and exertion of strength implied in manual labour . . . the more modern industry becomes developed, the more is the labour of men superseded by that of women. Differences of age and sex have no longer any distinctive social validity for the working class. All are instruments of labour, more or less expensive to use, according to their age and sex.[25]

This is a particularly important argument because it explains the way that capitalism increasingly turns women into wage-workers. Working class women and men form the "great camp" facing capital and it is thus increasingly important to the prospects for revolution that women conceive of themselves as working class. This is daily confirmed in the era of global capitalism, when women workers make up the increasing majority of global workers, subject to extremely low wages, and are particularly susceptible to the effects of austerity because they tend to work in and use the public sector more than men.

As my discussion thus far has implied, "new materialism" is a ruling class movement in cultural theory in general and in feminist theory in particular. "New materialism" is aimed not only at ideologically and pragmatically adjusting exploited workers to the exigencies of capitalism in crisis and marginalizing struggles for social transformation by representing them as outside the realm of the "sustainable" (as we see in Braidotti's theory of "new materialism"), but it also serves as a means to shore up the class privileges of a small ruling class minority of men and women in capitalism by translating class contradictions into a new metaphysics of freedom.

This is evident in the "new materialist" feminism of Elizabeth Grosz who, in her essay "Feminism, Materialism, and Freedom," calls for a turn in feminism to understanding freedom for women in "metaphysical terms"[26] and, more specifically, in terms of the autonomism of Nietzsche and the "vitalist" metaphysics of Henri Bergson. Grosz argues that feminism and its (so called) "traditional" concepts of materialism have been focused on a "negative" concept of freedom—what she calls a "freedom from"—by which she means freedom from external material relations outside the subject in which constraints are imposed by "the (oppressive and dominant) other."[27] Drawing on Nietzsche, Grosz characterizes struggles for "freedom from" as part of "the other-directedness of [a] reactive herd morality."[28] According to this narrative, "freedom from," therefore, only works "within existing social networks and relations" rather than "the creation of a future unlike the present."[29] While Grosz does not outright deny that constraints imposed by others still exist, she contends that "'freedom from' [. . .] at best [. . .] addresses and attempts to redress wrongs of the past without providing any positive direction for action in the future."[30]

Instead, Grosz argues that "we need to look outside the traditions of thought that have considered subjectivity as the realm of agency and freedom only through [. . .] the operation of forces—social, cultural, identificatory—outside the subject."[31] Feminism, she contends, should take up a "freedom to" which is "unconcerned with the other and its constraints, directed only to its own powers and to the fullest affirmation of its own forces."[32] "Freedom to," she claims, is a "positive" concept of freedom—a sign not of "herd morality" but of "the self-affirmation of an active or noble morality"—because it is not "bound up with a shared existence with the other and the other's powers over the subject"[33] and does not seek freedoms "bestowed on us by others," rather, it relies on "autonomy, in the individual's inner cohesion and historical continuity."[34]

Grosz's relegation of "freedom from" to an ignoble "herd morality" and the problems of the past is a ruling class caricature of collective struggles for social transformation that marginalizes the continued need for freedom from necessity brought on by the deepening of exploitation in transnational capitalism. The unsaid logic of the argument that collective struggles to realize freedom from necessity display a shameful lack of autonomy is the same ruling class moralism which asserts that the workers are seeking special favors and handouts from those in power instead of relying on their own "inner" resources and innovations. We can see the class politics of this theory when Grosz poses the rhetorical question: "Is feminist theory best served through its traditional focus on women's attainment of a freedom from patriarchal, racist, colonialist, heteronormative constraint? Or by exploring what the female—or feminist—subject is and is capable of making and doing?"[35] The underlying logic and unsaid assumption of Grosz's argument is that feminism no longer needs to concern itself with the emancipation of women from material relations and the social structures that oppress and exploit them in transnational capitalism. Contrary to Grosz's claims to supersede liberal feminism, her argument is not rooted in materialist or transformative feminism; rather, it is based on a ruling class postfeminism that implicitly assumes that women have already achieved "equality" and "power" and that struggles for collective social transformation of our "shared existence" are no longer relevant or necessary to bring about freedom for women.

This is a disenabling view for an emancipatory materialist feminism because it conceals the fact that despite the changes in women's lives that have come with being pulled en masse into the workforces of global capitalism, these changes have been class divided: while a small class fraction of women in global capitalism now, alongside of ruling class men, own a greater share of the wealth produced by the surplus-labor of the global working class and occupy positions of power, for the majority of women this has meant what Goretti Horgan calls "equality in exploitation" and the "race to the bottom upon which global capitalism is founded."[36] Close analysis of shifts in wages supports Horgan's argument. For instance, in "What Gender Does," Nancy Lindisfarne and Jonathan Neale note that in the United States since 1975, under neoliberalism, the closing wage gap between women and men (using the "median" as the benchmark) is an effect of declining men's wages rather than substantial progress for women workers in incomes, which have grown more slowly since 1975 than they did during the long post-war boom.[37]

As Horgan argues, social movements, including feminism, which fought for the liberation of women and other oppressed peoples, which necessitates the end of exploitation, have been displaced by forms of liberalism, including liberal feminism, which advocate for "equality" (with men) within the existing structures of capitalism. Grosz and other "new materialist feminists" are providing postfeminisms which are updated forms of such liberal feminism which displace materialist theory in feminism with patriarchy theories. Patriarchy theories see "the main divide in society as between women and men, rather than rich and poor. They [see] all men as having power over all women. This allow[s] them to argue that getting women into positions of power—managers, judges, politicians—[will] help all women."[38] Materialist theory, in contrast, argues that it was the historical emergence of the class divide that gave rise to gender inequality and that this class divide continues to necessitate gender as well as other forms of inequality. Women provide a source of cheap labor for capitalists as well as providing the vast majority of reproductive labor needed to reproduce the workforce from day to day and generation to generation, reproductive labor which is either un- or under-paid and justified as such by the ideology that naturalizes women as biologically suited to nurturing. Fighting for "equality" in relation to men within capitalism or, in Grosz's case, in what amounts to the same thing at a later historical moment, for (bourgeois) "freedom to" or "liberty" benefits ruling class women but will not end but rather extend and enable deepening exploitation and, ultimately, reinforce the oppression of working class women.

Take, for instance, the case of Walmart, which is the largest private employer in the United States. According to *Forbes* Magazine's "Forbes 400," the list of the 400 Richest People in America, forty-six of these people are women, and four of these women are members of the Walton family whose wealth comes from owning Walmart. These four women have a combined wealth of a staggering 71 billion dollars.[39] Yet, not only has Walmart resisted their employees' struggles to attain a $15.00 per hour minimum wage, finally acceding to only raise its base pay to $10.00 per hour by 2016,[40] but it has been fighting a sexual discrimination lawsuit in which "[a]lmost 2,000 women in 48 states claim that Walmart discriminated against them for pay and promotions."[41] At the same time, as CNN reports, in 2014 and 2015 Walmart made at least 16 billion in profits and "returned between $7 billion and $13 billion to its shareholders."[42] CNN goes on to report that the Waltons are worth 150 billion dollars, "which is more than the wealth of the entire bottom 40 percent of Americans. Walmart's owners, shareholders and executives have the hard work of its 1.3 million employees to thank for their incredible fortune."[43] The fact that a mainstream media outlet like CNN is compelled to criticize Walmart on the level of its exploitation of workers (though not, of course, on the system of exploitation Walmart exemplifies) marks the social crisis this obscene level of exploitation and resulting inequality has produced. Significantly, as Lindesfarne and Neale argue, "The greater the inequality in any particular place and time, the more gendered differences will be emphasized and enforced."[44] It is not surprising, then, that a 2003 study found pay inequity at all the status levels at Walmart—on average women workers made $5,200 less than men at Walmart.[45]

As an article on the website "Making Change at Walmart" argues, the case of Walmart is significant not because it is an isolated or extreme example, but because it is exemplary and, moreover, "[a]s the largest private employer in the world and the largest employer of women in the United States, Walmart sets the [standard] for how women are treated throughout the retail sector and the broader economy. In retail sales, women make just 68 cents for every dollar that men make and often face issues with pregnancy policy and problematic scheduling practices."[46] To argue for "equality" and an end of struggles for "freedom from the oppressive other" under these historical conditions is, in effect, to say that the problem for women and feminism is that there are only forty-six women on the Forbes 400 list rather than to address the structures of inequality that subject women workers in the United States and across the globe to such exploitation and oppression.

Grosz raises the historical and socially produced class contradictions in postfeminism to a new level of ahistorical abstraction by translating them into a new metaphysics of freedom. "Freedom," Grosz argues, is "an immanent and sometimes latent capacity in life."[47] "Life itself," and therefore "freedom," according to Grosz, is best explained as ontologically "indeterminate"—a radical form of contingency that "liberates life from the constraints of the present."[48] She argues that, "Indetermination is the 'true principle' of life, the condition for the open-ended action of living beings, the ways in which living bodies are mobilized for action that cannot be specified in advance"[49] and that "the possible never prefigures the real, it simply accompanies it as its post facto shadow."[50] "[E]ven in the most extreme cases of slavery," she contends, "there is always a small space for innovation [and . . .] inventiveness."[51] Freedom, in Grosz's narrative, is a spontaneous and inalienable condition of life itself or the sheer fact of living—an *élan vital* or spirit of life—such that even the slave has "freedom" within slavery, thus making the freedom from slavery unnecessary.

Grosz's "freedom to" or "within" which occults the conditions that necessitate freedom is a vacuous and empty concept of freedom that provides an apologetics for wage-slavery (and the growth of human trafficking) in capitalism. It puts in ideological suspension the historical and material conditions of necessity and exploitation which determine the ends and interests toward which "life" is put in capitalism by representing them as not only impossible but unnecessary to understand. In doing so, this theory is a most effective ally of capital because it rejects the possibility and necessity for collective social transformation, planning, and organization. Without the transformation of production relations, the "innovation" and "creativity" of the exploited—the capacity of the worker to produce—is still subordinated to production for the profit of some not the needs of all. Rather than becoming the basis of freedom for all, workers' surplus-labor enriches the owner's profit and conditions of life while workers are reduced to reproduction as exploited labor for the next working day. Grosz merely translates this process of alienation through exploitation into metaphysical terms and celebrates it as freedom.

Freedom is not an imaginary ontological autonomy but, as Engels argues, it is "a product of historical development."[52] Women and others can be free only insofar as they are free from necessity. Freedom from necessity is founded not on ontological autonomy but on social collectivity, which is a material relation of production in which no one person can privately own the means of production and, therefore, command over the surplus-labor—and thus the lives—of others; "a state of society in which there are no longer class distinctions or anxiety over subsistence for the individual, and in which for the first time there can be talk of real . . . freedom."[53]

Grosz is more concerned with the freedom of some from "definitions" than the freedom of all from exploitation. She writes: "Freedom has no given content; it cannot be defined" and explains this theory with the following statement from Bergson: "Any positive definition of freedom will ensure the victory of determinism."[54] However, Grosz does not actually leave freedom without a content—and, specifically, a class content. She writes: "Freedom is thus the exception rather than the rule in the sense that it can function only through the 'autonomy' of the living being against a background of routinized or habituated activity."[55] In other words, freedom is a matter of the "exceptional" living being who acts in unconventional ways. Yet, this libertarian "freedom" requires the continuation of "a background of routinized or habituated activity" that the so-called "exception" can be offset against. For the majority of women this means the normalization of their routine and habitual exploitation by capital which Grosz dehistoricizes by abstracting it from the material relations of production that enable it. Put another way, Grosz's theory ideologically normalizes and "disappears" the material exploitation of women's surplus-labor in transnational capitalism by ideologically "dissolving" it into a question of their "habituated activity." Grosz's "exceptionalism" puts into sharp relief the class interests behind her argument for a feminism that is "unconcerned with the other and its constraints" and that rejects "a shared existence with the other."[56] This is not a theory of the autonomy of the feminist subject from the "(oppressive or dominant) other"[57] as Grosz claims. Rather, it is a theory for a ruling class feminism which is aimed at dispensing with the collective needs of the exploited others in capitalism for freedom from necessity through collective social transformation. This is the mark of a turn to a capitalist class theory because capitalists benefit when we concern ourselves only with ourselves. When we do this (focus on the individual), we make ourselves as well as others easier targets for exploitation.

Despite its disclaimers, new materialism ideologically updates liberal feminism, and this not only leads it to reject conceptuality as the erasure of the random, the unpredictable, and the indeterminate, but to affirm the bourgeois theory of individual liberty, which underwrites new materialism's notions of affirmative ethics and autonomous freedom. Liberty, as Marx argues in "On the Jewish Question," is "[t]he right of the circumscribed individual, withdrawn into himself."[58] Individual liberty takes for granted a world in which individuals are divided by class: on the one hand there is the class of workers who are forced to compete with one another on the market for "jobs" in order to gain a livelihood and survive. On the other, there are the capitalists who buy the labor of workers on the market and profit from the use of that labor in production. What Marx calls the individual withdrawn into himself is thus the cultural effect of economic relations that make possible the individual who puts her own private need above collective need, and who thus accepts a world in which the have-nots struggle as individuals, increasingly alienated, in order to make ends meet. This is of course the individual whose liberty to remain an unregulated individual—to put the private before the collective—is lauded by the bourgeois, who take this right as a transhistorical, indeed natural phenomenon. As Marx puts it, the "practical application of the right of liberty is the right of private property."[59] Through the lens of individual liberty, capitalists' deepening of the rate of exploitation of workers may be criticized for being "too high" at times, when the class divide threatens to destabilize capitalism, but the ability of capitalists to exploit their workers is never questioned. Hence the emphasis, not on the need for transformation, but on the "selfless" and "noble" accommodation to worsening exploitation—a mantra that is everywhere repeated whenever working class movements demand better wages or working conditions.

A "materialism" that does not account for the historical relation of "life" in general or the conditions of women's lives in particular to the ensemble of the social relations of production—that is to class relations and exploitation in production—is not materialism but a new spiritualism that ideologically covers over the social structures and historical basis of inequality and economic contradictions in capitalism, and presents these as existential conditions of life as such. What Alexandra Kollontai said of bourgeois feminists of the twentieth century is still true today of "new materialist" feminists, including Grosz and Braidotti. She writes, "However apparently radical the demands of the [bourgeois] feminists, one must not lose sight of the fact that [they] cannot, on account of their class position, fight for that fundamental transformation of the contemporary economic and social structure of society without which the liberation of women cannot be complete."[60]

It is time to give up liberal, bourgeois theories, including "new materialist" feminism and take up historical materialist feminism for the twenty-first century.

BIBLIOGRAPHY

Braidotti, Rosi. "The Politics of 'Life Itself' and New Ways of Dying." In *New Materialisms: Ontology, Agency, and Politics,* edited by Diana Coole and Samantha Frost, 201–18. Durham, NC and London: Duke University Press, 2010.

Engels, Frederick. *Anti-Dühring.* Peking: Foreign Languages Press, 1976.

"Facts and Figures." *American Foundation for Suicide Prevention.* Last modified 2013. http://www.afsp.org/understanding-suicide/facts-and-figures.

"Forbes 400," *Forbes,* 2015. http://www.forbes.com/forbes-400/.

Frost, Samantha. "The Implications of the New Materialisms for Feminist Epistemology." In *Feminist Epistemology and Philosophy of Science: Power in Knowledge,* edited by H. E. Grasswick, 69–83. Dordrecht: Springer Science+Business Media, 2011.

Grosz, Elizabeth. "Feminism, Materialism, and Freedom." In *New Materialisms: Ontology, Agency, and Politics,* edited by Diana Coole and Samantha Frost, 139–57. Durham, NC and London: Duke University Press, 2010.

Hanson, Ellis. "The Future's Eve: Reparative Reading After Sedgwick," *South Atlantic Quarterly*, 110, no. 1 (2011): 101–19

Hines, Alice. "Walmart Sex Discrimination Claims Filed by 2,000 Women." *The Huffington Post*, June 6, 2012. http://www.huffingtonpost.com/2012/06/06/walmart-sex-discrimination-women-_n_1575859.html.

Horgan, Goretti. "How Does Globalization Affect Women?" *International Socialism* 92 (2001). http://pubs.socialistreviewindex.org.uk/isj92/horgan.htm.

Kollontai, Alexandra. "The Social Basis of the Woman Question." In *Selected Writings of Alexandra Kollontai*, translated by Alix Holt, 58–74. New York and London: W. W. Norton, 1977.

Lathrop, Yannett. "Don't Be Soothed by Walmart Pay Hike." *CNN.com*. February 25, 2015. http://www.cnn.com/2015/02/24/opinion/lathrop-walmart-wages.

Lenin, V. I.. *Materialism and Empirio-Criticism: Critical Comments on A Reactionary Philosophy*. New York: International Publishers, 1927.

Lindisfarne, Nancy, and Jonathan Neale. "What Gender Does." *International Socialism* 139, July 5, 2013. http://isj.org.uk/what-gender-does/.

Marx, Karl. *Economic and Philosophic Manuscripts of 1844*. In *Marx-Engels Collected Works*, vol. 3, 229–346. New York: International Publishers, 1975.

---. "On the Jewish Question." In *The Marx-Engels Reader*, edited by Robert C. Tucker, 26–52. New York: W. W. Norton, 1978.

---. *Wage-Labour and Capital*. In *Wage-Labour and Capital & Value, Price and Profit*. New York: International Publishers, 1976.

Marx, Karl and Frederick Engels. *The Communist Manifesto.* Edited by Frederic Bender. New York: W. W. Norton, 1988.

Mishel, Lawrence, Josh Bivens, Elise Gould, and Heidi Shierholz. *State of Working America*, 2012 ed., https://stateofworkingamerica.org/files/book/Chapter1-Overview .pdf

Moss, Michael, "The Extraordinary Science of Addictive Junk Food." *The New York Times*, February 20, 2013. http://www.nytimes.com/2013/02/24/magazine/the-extraordinary-science-of-junk-food.html.

"Productivity and Real Median Family Income Growth, 1948–2013." *State of Working America*. Last modified September 29, 2014. http://stateofworkingamerica.org/charts/productivity-and-real-median-family-income-growth-1947-2009.

"U.S.: Soaring Rates of Rape and Violence Against Women." *Human Rights Watch*. Last modified December 18, 2008. https://www.hrw.org/news/2008/12/18/us-soaring-rates-rape-and-violence-against-women.

Walmart. "Despite History of Pay Inequity at Walmart, Company Strikes Resolution From Proxy." Press release. *Making Change at Walmart*. Last modified April 23, 2015. http://makingchangeatwalmart.org/2015/04/23/despite-history-of-pay-inequity-at-walmart-company-strikes-resolution-from-proxy/.

Warren, Elizabeth, and Amelia Warren Tyagi. *The Two-Income Trap: Why Middle-Class Parents are Going Broke*. New York: Basic Books, 2003.

NOTES

1. Karl Marx, *Wage-Labour and Capital*, in *Wage-Labour and Capital & Value, Price and Profit* (New York: International Publishers, 1976), 44.

2. Samantha Frost, "The Implications of the New Materialism for Feminist Epistemology," in *Feminist Epistemology and Philosophy of Science: Power in Knowledge*, ed. H. E. Grasswick (Springer Science+Business Media, 2011), 70.

3. Ellis Hanson, "The Future's Eve: Reparative Reading After Sedgwick," *South Atlantic Quarterly* 110, no. 1 (2011), 105.

4. Rosi Braidotti, "The Politics of 'Life Itself' and New Ways of Dying," in *New Materialisms: Ontology, Agency, and Politics*, ed. Diana Coole and Samantha Frost (Durham, NC and London: Duke University Press, 2010), 210.

5. Ibid., 210.
6. V. I. Lenin, *Materialism and Empirio-Criticism* (New York: International Publishers, 1927), 348.
7. Michael Moss, "The Extraordinary Science of Addictive Junk Food," *The New York Times*, February 20, 2013, http://www.nytimes.com/2013/02/24/magazine/the-extraordinary-science-of-junk-food.html.
8. Karl Marx, *Economic and Philosophic Manuscripts of 1844*, in *Marx-Engels Collected Works*, vol. 3 (New York: International Publishers, 1975), 300.
9. Ibid., 302.
10. Ibid., 303.
11. Braidotti, "Politics," 208.
12. Ibid., 213–14.
13. Ibid., 210.
14. Ibid., 210.
15. Ibid., 211.
16. Lawrence Mishel, Josh Bivens, Elise Gould, and Heidi Shierholz, *State of Working America*, 2012 ed., 28, https://stateofworkingamerica.org/files/book/Chapter1-Overview.pdf.
17. "Productivity and Real Median Family Income Growth, 1948–2013," *State of Working America*, last modified September 29, 2014, http://stateofworkingamerica.org/charts/productivity-and-real-median-family-income-growth-1947–2009.
18. Mishel et al., *State of Working America*, 30–31.
19. Elizabeth Warren and Amelia Warren Tyagi, *The Two-Income Trap* (New York: Basic Books, 2003), 57–63.
20. Ibid., 64.
21. Ibid., 58.
22. Ibid., 58.
23. "U. S.: Soaring Rates of Rape and Violence Against Women," *Human Rights Watch*, Last modified December 18, 2008, https://www.hrw.org/news/2008/12/18/us-soaring-rates-rape-and-violence-against-women.
24. "Facts and Figures," *American Foundation for Suicide Prevention*, last modified 2013, http://www.afsp.org/understanding-suicide/facts-and-figures.
25. Marx and Engels, *The Communist Manifesto*, ed. Frederic Bender. (New York: W. W. Norton, 1988), 62.
26. Elizabeth Grosz, "Feminism, Materialism, and Freedom." In *New Materialisms: Ontology, Agency, and Politics*, eds. Diana Coole and Samantha Frost (Durham, NC and London: Duke University Press, 2010), 140.
27. Ibid., 141.
28. Ibid., 141.
29. Ibid., 154.
30. Ibid., 141, emphasis added.
31. Ibid., 140.
32. Ibid., 141, emphasis added.
33. Ibid., 141.
34. Ibid., 153.
35. Ibid., 141.
36. Goretti Horgan, "How Does Globalization Affect Women?" *International Socialism* 92 (2001), http://pubs.socialistreviewindex.org.uk/isj92/horgan.htm.
37. Nancy Lindisfarne and Jonathan Neale, "What Gender Does," *International Socialism* 139 (July 5, 2013), http://isj.org.uk/what-gender-does/.
38. Ibid., np.
39. "Forbes 400," *Forbes*, 2015, http://www.forbes.com/forbes-400/.
40. Lathrop, Yannett. "Don't Be Soothed by Walmart Pay Hike," *CNN.com*, February 25, 2015, http://www.cnn.com/2015/02/24/opinion/lathrop-walmart-wages.

41. Alice Hines, "Walmart Sex Discrimination Claims Filed by 2,000 Women," *The Huffington Post*, last modified June 6, 2012, http://www.huffingtonpost.com/2012/06/06/walmart-sex-discrimination-women-_n_1575859.html.
42. Lathrop, "Don't Be Soothed."
43. Ibid.,
44. Lindisfarne and Neale, "What Gender Does."
45. "Despite History of Pay Inequity at Walmart, Company Strikes Resolution From Proxy," *Making Change at Walmart*, last modified April 23, 2015, http://makingchangeatwalmart.org/2015/04/23/despite-history-of-pay-inequity-at-walmart-company-strikes-resolution-from-proxy/.
46. "Despite History."
47. Grosz, "Feminism," 149.
48. Ibid., 153.
49. Ibid., 149.
50. Ibid., 147.
51. Ibid., 154, note 1.
52. Friedrich Engels, *Anti-Dühring*, (Peking: Foreign Languages Press, 1976), 144.
53. Ibid., 145.
54. Henri Bergson quoted in Grosz, "Feminism," 147.
55. Grosz, "Feminism," 148.
56. Ibid., 141.
57. Ibid., 139, emphasis added.
58. Karl Marx, "On the Jewish Question," in *The Marx-Engels Reader*, ed. Robert C. Tucker. (New York: W. W. Norton, 1978). 42.
59. Ibid., 42.
60. Alexandra Kollontai, "The Social Basis of the Woman Question," in *Selected Writings of Alexandra Kollontai*, trans. Alix Holt. (New York and London: W. W. Norton, 1977). 59–60.

FIVE

Ghostly Objectivity

Commodity Fetishism, Animated Monsters, and the Posthuman Object

Rob Wilkie

We live in an epoch of dead ideas, in which the North Atlantic left is "anxiously [conjuring] up the spirits of the past . . . to present the new scene of history in this time-honoured disguise and this borrowed language."[1] As the capitalist nations spend the first part of the twenty-first century like the beginning of the last, lurching blindly from one crisis and conflict to the next, the inability of capitalism to address the needs of the world's population much less meet its own immanent goals of sustained rates of growth and profitability is reflected in the alchemic attempts by "posthumanist" theorists to give "the breath of life"[2] to nonhuman objects in order to naturalize the alienation of capitalist society as the ontology beyond which no transformation is possible.

Left Fetishism and the "Object-Turn"

Speaking in a "left" discourse that long ago decided concepts like class, critique, and exploitation were "zombie categories"[3] that need to be abandoned, writers such as Graham Harman (*The Quadruple Object*), Levi Bryant (*The Democracy of Objects*), Jane Bennett (*Vibrant Matter*), Ian Bogost (*Alien Phenomenology*), Timothy Morton (*Hyperobjects*), Quentin Meillassoux (*Against Finitude*), and Tristan Garcia (*Form and Object*) argue that capitalism's epiphenomenal series of social, political, economic and environmental crises are the outcome of a cultural logic of human instrumentality rather than class exploitation. The problem of the contemporary is thus not capitalism *per se* but human-centered theories of ontology that crisscross economic and political formations and privilege human beings over all nonhumans—from animals (Simondon, *Two Lessons on Animal and Man*) to rocks (Cohen, *Stone: An Ecology of the Inhuman*)—without recognizing the "agency" of the nonhuman in its own right.

Western theories of ontology, it is said, effectively marginalize both the human and the non-human by treating the world as if it consists of "life-less" objects just waiting to be put to use by humanity, while subsequently reducing objects to their existence as defined by human relations. What the proponents of the object-turn argue is that all social formations, whether capitalist or communist, inherently focus on the status of "human" agency without accounting for the agency of nonhumans and are therefore equally at fault, regardless of their social priorities. As Jane Bennett writes, "the image of dead or thoroughly instrumentalized matter feeds human hubris and our earth-destroying fantasies of conquest . . . by preventing us from detecting (seeing, hearing, smelling, tasting, feeling) a fuller range of the nonhuman powers circulating around and within human bodies."[4] To claim to "know" anything about reality is to eliminate its multiplicity and to strangle "life" with a reductive epistemology. According to Bennett, this means conceiving of objects not as "intrinsically inanimate matter"[5] but "as vivid entities not entirely reducible to the contexts in which (human) subjects set them, never entirely exhausted by their semiotics."[6] The "thing" or "object" is celebrated as an independent entity with its own "being," irreducible to structural relations (human or otherwise) of any kind. If every object—human and nonhuman alike—exists not only in isolation from each other but also independently from any relation, all that can be accomplished in terms of changing the relationship between the human and the nonhuman is an ethical encounter-by-encounter recognition of the limits of the human intellect to understand all of the ways in which the nonhuman objects act on their own, autonomously from human intention, awareness, or understanding.

The "ethical" turn to the object is, as all North Atlantic left theory tries to claim today, a supposedly postbinary logic that undoes what William Connolly dismisses as "eliminative materialist" readings of the social which are "eliminative" in the sense of defining "reality" by eliminating any possibility of nonhuman agency outside of human relations. Despite assertions that the turn to the object represents a "shift from the language of epistemology to that of ontology,"[7] this is a change in thought that remains focused on representations rather than the material inequalities of capitalist society. The "thing" is no longer "material" but "real" and becomes in this discourse what Derrida calls the trace of *différance*,[8] threatening to rupture all epistemological foundations including the reduction or reification of instrumental logic. To be ethical, in this context, means to "dissipate the onto-theological binaries of life/matter, human/

animal, will/determination, and organic/inorganic"[9] by recognizing objects as existing as "quasi-agents or forces with trajectories, propensities, or tendencies of their own" beyond the limits of human intentionality and understanding.[10] The embrace of "things" or "objects" or "being" outside of social relations is to render it impossible to understand the ways in which social relations shape our understandings of other beings. In fact, to suggest that human relations can explain social crises in any way—whether economic, environmental, social, or political—would be to assume that such problems might not be the result of nonhuman objects acting on their own, outside of human consciousness. On these terms, there is no basis upon which to claim that capitalism has anything to do with the contradictions of contemporary life, as financial crises or environmental catastrophes may just as well be caused by nonhuman agents operating beyond our ability to conceptually grasp their actions.

Yet, even as the proponents of the object turn displace epistemology with calls for a nonrelational ontology, the only kind of "resolution" to the conflict between "humans" and "things" that is offered is one that takes place at the level of ideas, through the renewal of an ethical sense of the "vitalism" of objects that deprioritizes and decenters the human by acknowledging the object's "lively" status as an equally active agent in the world. It begins with the assumption that the division between the human and the nonhuman is established by a post-Kantian thinking that pivots around the human capacity to know that which is nonhuman. It then proceeds from this division between the human and the nonhuman object to offer solutions that amount to little more than doing away with post-Kantian binaries without accounting for the ways in which relations between humans and objects are determined historically, by which I mean looking outside of epistemology in the mode of production in order to explain why cultural logics such as humanism or posthumanism emerge when they do. In this way, the "ecological" disguise of "[extending] our political and cosmic sensibilities" in order to embrace "a variety of nonhuman force fields that impinge upon politico-economic life" in ways that are "irreducible simultaneously to eliminative materialism, to mechanistic theories of causation, to grand teleologies pulled by a final purpose, to simple human intentionalism, to the most familiar notions of progressive time, to any notion of complete explanation, and to the sharpest lines of division between nonliving nature and human agency"[11] actively *obstructs* from understanding the historicity of the object, by which I mean, as Marx writes, "the orientation to the object [as] the *manifestation of the human reality*."[12] Just as Marx demonstrates throughout his work that the "phenomenal" characteristics of the world around us, as well as all theories of "self-hood," are a reflection of the historical and material conditions of production, the sudden interest in "objecthood" is a distorted expression of the deep alienation of labor that exists in contemporary capitalism. To approach reality in the manner of the proponents of the object-turn is to proceed from a surface level reading of the existing and, rather than look for causes, to project the existing conditions indefinitely into the past and the future.

In contrast to such an approach, as Marx writes,

> It is not the *unity* of living and active humanity with the natural, inorganic conditions of their metabolic exchange with nature, and hence their appropriation of nature, which requires explanation or is the result of a historic process, but rather the *separation* between these inorganic conditions of human existence and this active existence, a separation which is completely posited only in the relation of wage labour and capital.[13]

If a stone becomes a flint axe, the foundation of an office building, an aesthetic element of a cultural practice like an engagement ring, or a weapon in a Palestinian slingshot, for example, it is not because "there is something in rock that is actively unknowable"[14] but the mode of production which shapes what labor can make of the stone and what ends that stone will be put to use. Material relations make "things" possible, but the way in which objects are reified and come to appear autonomous from historical relations—as if they have an *élan vital* or metaphysical life force of their own—is the manifestation of the extent to which all elements of the world around us have been subsumed within the capitalist mode of production. The nonrelational ontology of the object-turn—which addresses the contradictions between humans and things by, on the one hand, denying that any such contradiction exists outside of epistemology and, on the other hand, by claiming that *all* objects are equally unknowable and divided from every other object in an extreme form of autonomous isolationism—reproduces the material conditions of capitalism in which, as a result of the private ownership of the means of production, "living labour appears as a mere means to realize objectified, dead labour, to penetrate it with an animating soul while losing its own soul to it."[15] Thus, while the majority of the world's population are reduced by wage-labor to the status of commodities on the market, the posthumanist object-turn argues that objects cannot be reduced to their position in social relations. According to this logic, if you are treated like an object under capitalism, as people are when they are dependent on the sale of their labor-power for survival, what is necessary is not changing the social relations that create such conditions of exploitation, since all relations operate in essentially the same ways by reducing the other to a false epistemology, but realizing that "you" are not reducible to any of them. In fact, the reification of wage-labor in capitalist society becomes the basis for an existential individualism in which humans and things alike can find "freedom" by embracing the ultimate unknowability, contingency, and emptiness that the majority of people face living under capitalism.

The object-turn is a defense of capitalism which shifts theory away from the revolutionary praxis of red critique and towards a descriptive postcritique mode of writing Jane Bennett calls "a childhood sense of the world"[16] and Steven Shaviro describes as "a kind of 'autistic' thought"[17] which seeks "not to deduce and impose cognitive norms or concepts of understanding, but rather to make us formally aware of how reality escapes and upsets these norms."[18] If objects take on a "ghostly objectivity" wherein they seem to exist as autonomous agents above and beyond the human—as it appears with seemingly "nonhuman" agents like stock markets, drones, or climate change—it is because the process of commodification in which the products of human labor are stripped from the worker creates "an autonomy that seems so strictly rational and all-embracing as to conceal every trace of its fundamental nature: the relation between people."[19] Capitalism is the *historical* reality that the object-turn reifies and renders as *unhistorical* and therefore *untransformable* ontology.

What is particularly telling is just how comfortable global capitalism is with the anti-instrumentality that object-turn proponents claim is resistant to all cultural logics. The 2014 big-budget Hollywood "reboot" of *Godzilla*, for example, firmly locates the cause of contemporary crises in the fundamentally flawed logic of human instrumentality and through the actions of a nonhuman other "exposes" the limits of human understanding. Like left theory's "ghostly objectivity," *Godzilla* expresses a similar inability of capitalist ideology to address the crises of the contemporary by, first, turning the contradictions of capitalism into the trope of a conflict between giant prehistoric kaijū ("monster"), and, second, representing these contradictions as operating beyond human control and understanding. Both "left" theory and popular culture today work to absolve the contradictions of capitalism by eliminating all investigation of "relations" and, through the figure of the nonhuman object, placing the blame on an abstract and empty notion of "humanity" while looking to an equally empty and ahistorical notion of the nonhuman object for solutions. The idea that "the true principle of dynamism, in human society as well as inanimate nature, is that real objects always exceed their contexts, always withdraw from our control, and are always filled with surplus and surprise"[20] is the ideology of commodity fetishism which erases the fundamental role of labor in the production of surplus value and obscures the possibilities of transformation by representing the historical relations between people as "the fantastic form of a relation between things."[21] The object-turn is now taking its turn in the spotlight of ruling class ideology because, as is the case with all ideology, it dispenses with the notion that exploitative social relations can in any way account for the contradictions that result from production for profit. Instead, it places any "blame" on unforeseen and unknowable events. It does so not by providing an alternative explanation of social contradictions, but by declaring that ontology exceeds explanation and cannot be reduced to historical relations. In proposing that "entities are never exhausted by any of their relations or even by their sum of all possible relations" and "no entity is capable of fully registering the depths of another,"[22] the object-turn fosters the illusion that the problems facing society are not reducible to capitalism since it claims the problem of nonrecognition exists within all relations. This is to turn the effects of capitalist property relations—in which individuals become "autonomous" beings who see themselves as disconnected from the social by the very fact that they must compete against others to sell their labor on the market—into the condition of being *as such*. It is a theory of ontology that turns the *effects* of historical (class) conflicts into the *actions* of wayward objects. In this way, the object-turn acts as the theoretical arm of advanced capitalism in promoting the idea that social crises are the consequence of the "regulations" that are placed on life and the forms it can take, and which must be removed for life to flourish. Returning the conditions of labor to discussions of the object is the means by which we can understand why the object has become a theoretical "problem" and turn toward an examination of the world around us in a way that enables fundamental social transformation from an exploitative society to one focused on meeting human needs.

Heidegger's Empty "Thinking"

Despite immanent differences with one another, the various representatives of the object-turn take as their starting point the phenomenological challenge to instrumentality advanced in the work of Kant, Husserl, Merleau-Ponty, and Heidegger, and seek the solution to what is represented as a crisis of thought in a postmaterialist "realism" beyond the human. According to the proponents of the object-turn, traditional theories of ontology, particularly since Kant, are "correlationist" in that they assume that the ability to discern causes and effects lies entirely in human thought and thus, in privileging human cognition, fail to account for the ways in which "our day-to-day, spiritual, emotional, sexual, social, cultural and political lives are conducted in relation to objects and thoroughly mediated by them in whatever forms they take, qualities they possess, and complex practices they enable."[23] According to this narrative, to recognize the central role that objects have in shaping human life means, on the one hand, "one must abandon the belief that human access sits at the center of being, organizing and regulating it like an ontological watchmaker"[24] and, instead, realize that "events (a blackout, a meal, an imprisonment in chains, an experience of litter)" represent "encounters between ontologically diverse actants, some human, some not, though all thoroughly material."[25] On the other hand, accepting the "agency" of objects means also accepting that "things lie beyond the grid of intelligibility"[26] and that it is impossible to devise a theoretical approach which could "arrange the universe of all things according to an eminent object which would comprehend them all together."[27] Objects, in other words, are the singularity that exceeds all relations and, instead, interact through momentary and contingent "encounters" without any overarching structure. Since objects resist all attempts at reducing them to a relational structure, to embrace the object on its own terms requires an aesthetic "feeling"—not classifying, critiquing, nor even being able to understand—the diversity and expansive nature of *all* objects, placing them in Manuel DeLanda calls a "flat ontology" of "interacting parts and emergent wholes . . . made exclusively of unique, singular individuals, differing in spatio-temporal scale, but not in ontological status."[28] Putting "*things* at the center of being," as the flat ontology of the object-turn does, results in the belief that "nothing has special status, but that everything exists equally—plumbers, cotton, bonobos, DVD players, and sandstone."[29]

This is *anti-humanist* philosophy masquerading as what Levi Bryant calls the "democracy of objects."[30] The use of terms like "human hubris"[31] and "human intentionality"[32] might appeal to certain "left" sensibilities in the sense that they lend the appearance of caring about (all) things at a time of heightening social crises, but they are utterly meaningless. Does humanity *inevitably* suffer from "hubris"? Is "intentionality" the logic of *all* human societies? If so, this means that the only resolution to the opposition between the "human" and the "object" is the elimination of the human. Of course, the proponents of the object-turn would claim that the "human" is itself a product of "human hubris" and "human intentionality" insofar as the "human" is nothing but an epistemological creation of Western thought. Thus eliminating the "human" does not mean eliminating humanity but equalizing the relations between "humans" and "things" by deconstructing any concepts that create a binary between the two. However, stating that the human is a product of thought is nothing but a tautology (the human creates the epistemology that creates the human). Without any recourse to an historical outside to epistemology, there is no accounting for why Western conceptions of the "human" and the "object" emerge in the first place. "Hubris" or "intentionality" become either the consequence of intellectual accident or the inevitable outcome of the human mind. In either case, if there is no outside to the epistemological opposition between the "human" and the "object" from which to understand its very existence, then there is no clear basis upon which to claim society should change or what it should change to. In placing the blame on a self-producing concept of the "human," the proponents of the object-turn are left chasing the "ghostly objectivity" of a thoroughly dematerialized and therefore static reality in which the empty abstractions of "human" and "object" collide without clear cause, consequence, or resolution. They erase the fact that if humans and things appear to be "equal" today it is because capitalism has turned the majority of humanity into commodities who must compete with each other, as well as with the very products of their labor in the form of technology, on the market in order to secure a job.

The *arche*-theorist for the object-turn is Martin Heidegger, who continues to live on in North Atlantic left theory regardless of his history of fascism and anti-left thinking (e.g., Derrida, *Of Spirit: Heidegger and the Question*). Heidegger's work remains influential because of the way in which it couches a postclass and postcritique mysticism in an "ecological" and "anti-instrumental" discourse that appeals to a left sensibility concerned with the effects of capitalism while simultaneously dematerializing and dehistoricizing capitalism's contradictions in both style and substance such that nothing about reality ever has to actually change. In both his central philosophical work *Being and Time* and his later essay "The Thing," Heidegger turns to the question of the being of objects in order to challenge what he argues is the way in which Western philosophy "presupposes" the world, reducing "objects" (and later "things") to ontical concreteness that obscures a deeper ontological consciousness, and results in the instrumental and scientific logic in which "everything everywhere is ordered to stand by, to be immediately on hand, indeed to stand there just so that it may be on call for a further ordering."[33] The inability to leave aside interpretive tendencies and presuppositions, Heidegger argues, is reflected in the growing embrace of instrumentality, science, and technological advances that usher in the promise of eliminating "all distances in time and space."[34] Writing at the beginning of the twentieth century during a period of heightened social contradictions—resulting, on the one hand, in two global capitalist wars and, on the other, the dialectical negation of capitalism which emerges with the material possibility of socialism—Heidegger could not ignore these contradictions, but instead attributes the resulting alienation and conflict of capitalist society to ideas and seeks refuge in a static and hierarchical reality that resists social transformation. He claims the problem is not capitalism, but the ways in which Western instrumentality creates an "unsettling" existential terror as a result of the fact that "the frantic abolition of distances brings no nearness."[35] According to Heidegger, the embrace of human intentionality in scientific and technological progress in transportation, industry, and telecommunications generates the illusion of a deeper understanding of reality while actually reducing our understanding of the world, and by extension, of *Dasein*. He argues that Western philosophy approaches the question of ontology from within a framework that "always encounters only what *its* kind of representation has admitted beforehand as an object."[36] On the contrary, Heidegger writes that what is necessary for "the achieving of phenomenological access to entities which we encounter" is to "[thrust] aside our interpretive tendencies, which keep thrusting themselves upon us and running along with us, and which conceal not only the phenomenon of such 'concern,' but even move those entities themselves *as* encountered of their own accord *in* our concern with them."[37] In other words, what is necessary to bring ourselves closer to the "neighborhood" of truth is to move beyond interpretive presuppositions. We must reject science and critique while embracing a mode of poetic "nonthinking" that takes "the step back from thinking that merely represents—that is, explains—to the thinking that responds and recalls."[38]

In particular, what Heidegger finds most "unsettling" throughout his work is the attribution of any causality or agency to human labor, as this could potentially lead to a questioning of the conditions of labor rather than just "correcting" the (mis)recognition of objects. In "The Thing" Heidegger writes "when and in what way do things appear as things? They do not appear *by means of* human making."[39] More specifically, when talking about the existence of a jug he says that "the making, it is true, lets the jug come into its own. But that which in the jug's nature is its own is never brought about by its making . . . the jug's thingness resides in its being *qua* vessel."[40] Heidegger then finds that the being of the jug "does not lie at all in the material of which it consists, but in the void that holds."[41] In focusing on the "void" that the jug "reveals" in its making, Heidegger dematerializes the jug and dehistoricizes the labor which produces it, claiming in an equally immaterial discourse that "the world's worlding cannot be explained by anything else nor can it be fathomed through anything else . . . causes and grounds remain unsuitable for the world's worlding."[42] If labor is "productive," it is not in the sense of value or even meeting needs, but in bringing humanity closer to revealing the truth of being. For Heidegger, as well as his intellectual descendants, with the revealing of the void we become aware that the truth of the world rests on no material basis, or at least any determinate or dialectical understanding of causes and effects. The "meaning" or "cause" is literally absent in the jug's very purpose *qua* vessel: to be empty in order to hold something else. To "reveal" the world—what he calls the "world's worlding"—can only be accomplished through circumscription in spontaneous but natural relations between humans and objects.

Despite changing his use of the term "thing" in its later work from dismissal to appreciation, he advances a similar argument in the earlier *Being and Time* when, in discussing the "thingness" of a hammer, he distinguishes between the "present-at-hand," or that which something seems on the surface to the empirical or scientific consciousness, and that which is "ready-to-hand," or that which relates and appropriates the thing "in a way which could not possibly be more suitable."[43] Just as the jug maker does not contemplate the "void" or immaterial realism in advance of the making of the jug but instead inadvertently reveals that which is beyond the "present-at-hand" in the process, "the less we just stare at the hammer-Thing, and the more we seize hold of it and use it, the more primordial does our relation to it become."[44] Heidegger's dismissive attitude toward labor here extends to the intentionality of the laborer to the tool. Heidegger states:

> When we concern ourselves with something, the entities which are most closely ready-to-hand may be met as something unusable, not properly adapted for the use we have decided upon. The tool turns out to be damaged, or the material unsuitable. In each of these cases *equipment* is here, ready-to-hand.[45]

Heidegger proposes that the instrumental logic of "present-at-hand" means that we often confront the true "ready-to-hand" of the hammer when it *doesn't* work or is missing. Although we might blame the tool for our failure, it is not the tool, but the intention that is at issue as the tool is still "ready-to-hand." The problem of the "un-ready-to-hand" is that it cannot be grasped through the ontic, or "by looking at it and establishing its properties."[46] He writes, "anything which is un-ready-to-hand in this way is disturbing to us" and it is when our intentions are "disturbed" that we can see the intentionality with which we initially and unconsciously approached the object. Heidegger argues,

> when something ready-to-hand is found missing, though its everyday presence [Zugegensein] has been so obvious that we have never taken any notice of it, this makes a *break* in those referential contexts which circumspection discovers. Our circumspection comes up against emptiness, and now sees for the first time *what* the missing article was ready-to-hand *with* and *what* it was ready-to-hand *for*. The environment announces itself afresh.[47]

This "emptiness," like the jug's exposure of the "void," represents that which is in excess of human understanding and intention. It means that the object does not "present" itself, but rather in withdrawing from human consciousness *destructs* the absent cause upon which all "present-at-hand" thinking, or the instrumentality at the heart of Western thought, presupposes as its justification.

It is this reading of the object as most "truthful" in its withdrawal from the human that has become, despite Heidegger's ultimate insistence on the being specific to humanity (*Dasein*) as the only being able to truly grasp the ontological "truths" of the world, the basis for the contemporary object-turn. What this means in practice is the declaration that the only thing we can know about things is that we know nothing about them. Jeffery Jerome Cohen, for example, writes, "every object holds unfathomable reserves and cannot be equated to anything else as a way of depleting its possibilities,"[48] while Boria Sax declares "our abstract categories do not apply very well to the natural world, for they were created to describe the world of men and women."[49] Similarly, Steven Shaviro argues, "knowledge is just one particular sort of relation—and not even an especially important one at that. Most of the time, entities affect other entities blindly, without knowledge playing a part at all,"[50] and Ray Brassier concludes that "Thinking has interests that do not coincide with those of the living; indeed, they can and have been pitted against the latter."[51] Ultimately, to grasp the object in its withdrawal from thought means that any sort of "demystifying" critique which assumes the ability to discern causes from effects must go. As Jane Bennett contends, critique presumes that we must be "on high alert for secret signs of truth (a human will to power) below the false appearance of nonhuman agency."[52] The central "debate" amongst the proponents of the object-turn turns out to be who can outdo the other in declaring how little we can ever know or understand about the world around us. What they make a problem of epistemology, however, is a historical problem. That human knowledge is not static and that the relations between the human and the nonhuman change is not meaningful in itself, since neither confronts the deeper issue of *why* such changes take place. For instance, why, on their own terms, should there suddenly be an outpouring of interest and concern for the nonhuman? And, further, why on these particular terms and not others? As such, what is presented as an epistemological failure cannot be not resolved through the object-turn's declaration that historical analysis is unable to fully grasp the singularity and autonomy of the object. In fact, the idea of the object's autonomy from any relations (historical, social, environmental . . .) perpetuates the market logic of capitalism. What other social system preaches absolute independence from others as the highest form of "freedom" except one that depends upon the fact that the majority of people are "free" from any means to survive except the sale of themselves in the form of their labor-power?

"Object-oriented" philosopher Graham Harman is perhaps the main theorist of the object-turn's "noncritical theory"—or what he advocates as "naiveté."[53] His reading of Heidegger's theory of "tool-being" has become the bedrock of the object-turn, in part for how it gives Heidegger's rigid and static reality a "pluralist" sheen. Calling himself an "anti-materialist" realist,[54] Harman argues "direct knowledge of anything is impossible because to be truly direct, knowledge of a thing would have to be that thing itself"[55] when, in fact, "each thing is an inexhaustible surplus."[56] For instance, regardless of how many descriptions or accounts we might generate about a tree, human cognition ("ideas") can never take the place of the tree ("being"). His "contribution" to Heideggerian thinking is that unlike Heidegger's focus on *Dasein*, Harman claims the tree is always withdrawing from not just from human cognition, but also from all other objects such that no object can "know" it. Writing that "the basic principle of object-oriented philosophy . . . is that objects withdraw from all theoretical and practical contact alike"[57] he argues that the limits of all prior phenomenological approaches from Kant to Husserl to Heidegger is that they nonetheless remain too human-centered, even as they seek to uncover the limits of knowing objective reality beyond the boundaries of human intellect. Instead, Harman claims to go further in dismissing materiality than any prior "realist" by beginning with the principle that any "model of a world having reality outside the mind, a world in which some things are connected but not everything, in which things have an essential inner reality but not one that can be known or mastered, and can only be approached obliquely through allusion rather than directly through catalogues of tangible properties"[58] has to be extended to include objects as well. In other words, not only humans, but things too, know nothing about the world in which they live or the beings that inhabit that world. No "being" knows anything about any "thing," nor does any "thing" know anything about "being."

Taking as his starting point Heidegger's analysis of "tool-being" in *Being and Time,* Harman proposes that Heidegger's real contribution to ontology is not only attacking instrumentality but discovering that "if tool-beings are worthy of greater esteem than the images in consciousness, this is not because they are more dependent on human *Dasein*, but the opposite."[59] In other words, Harman takes the "void" or "emptiness" and uses it to diminish further human agency and intellect, both of which depend upon the real target, namely human labor. The "weird realism"[60] of the nonmaterial real, according to Harman, is that "even if we perceive every quality of an object perfectly, we would still not reconstruct the thing in its reality,"[61] because "the object is a dark crystal veiled in a private vacuum: irreducible to its own pieces, and equally irreducible to its outward relations with other things."[62] As an example he describes the way in which "when fire burns cotton, it makes contact only with the flammability of this material" and that "the being of the cotton withdraws from the flames, even if it is consumed and destroyed."[63] In other words, it is not only that "reality slips from view into a perpetually veiled underworld, leaving [humans] with only the most frivolous simulacra of these entities,"[64] but that "the withdrawal of objects is not some cognitive trauma that affects only humans and a few smart animals, but expresses the permanent inadequacy of any relation at all."[65] What this means for Harman is that there is a fundamental singularity and autonomy to *all* objects—human and nonhuman—that can never be reduced to relations or essences because all relations, essences, and objects exceed any understanding or connection to another. Everything is a surplus that exceeds every epistemology, rendering all being fundamentally unknowable. In turn, the materiality of objects—their existence in relation to one another—is denied as superfluous to their existence as objects. "Their reality," he writes, "consists in solely being what they are, not in some sort of impact on other things."[66] In eliminating the ways in which the interactions of objects shapes their existence, Harman turns the question of being into the renunciation of causality, and thus the opposite of science. Despite claiming that it would be difficult to construct an "object-oriented theology" because "an omniscient entity would not just need to know all things, but *to be* all things,"[67] Harman's vision of nonrelational objects whose "being" is dependent upon nothing but that they "exist" is a warmed-over spiritualist account of reality as driven by spontaneous generation reminiscent of pre-seventeenth century idealist science and contemporary anti-evolutionary preachers. By proclaiming that material relations have no bearing on the being of the object, Harman presents as a theory of "autonomy" and "agency" the illusion of a static world that resists transformation.

In representing objects as nonrelational, Harman seeks to eliminate critique from contemporary thinking by painting the picture of a world in which "*not everything is connected.*"[68] For instance, in describing how his "naïve" approach would address an issue like global warming he argues that "a holistic approach" which assumes "people and climate are not separate, but interconnected" is "the wrong one to draw from climate change" because it "weakens our ability to anticipate possible *surprises* to our climate models."[69] According to Harman, a world that is fundamentally indifferent and impossible to know in any kind of totality requires "a buffer zone of uncertainty" that includes "chance fluctuations in the activities of earthquakes, volcanoes, icebergs, or monkeys no less than conscious human action" and means that "historical determinism [is] impossible."[70] For instance, perhaps practices we think might be harming the environment "will turn out to be less damaging than expected" while at the same time "maybe some simple consumer chemical thought to be harmless, or perhaps a gas emitted from dying beetles, turn out to be catastrophic factors in the climate process."[71] This "openness" to possibility is the perfect corporate logic and mirrors in the way that U.S. business leaders and politicians attempt to displace any questioning of the connection between capitalism and environmental destruction by claiming that they are "not scientists" but that science knows nothing definitive anyway.[72] Like them, Harman declares that "most of any situation is unstated and unstatable, and is best approached by way of hint, allusion, oblique reference, and innuendo, just as we do in everyday language."[73] Who is to say, in other words, whether capitalism, dead beetles, or some yet unknown object is the "reality" of climate change that is hiding forever in a singular realm beyond human consciousness? History, he concludes, "is less a linear progression from dark to light (as revolutionaries think), or even from light to dark (as reactionaries think), but a dizzying series of reversals or retrievals that convert dead forms into new and living media."[74] This is analysis of the social as a game of "blind man's bluff." The element of surprise is not a sign of openness to opportunity and change, as the "object-oriented" thinkers claim, but rather the means by which ideology forecloses revolutionary change by turning the anarchy and uncertainty of life under capitalism into a first principle. This is why human knowledge is foreclosed in advance as little more than "hubris" because it means nothing can be done but learning to live with the "inevitable" crises, catastrophes, and the uncertainties of life. Of course, this does not mean that "uncertainties" do not happen, but rather that even uncertainty is always historical. How humanity deals with the "uncertain" is fundamentally a class issue. An earthquake or volcano eruption might be unexpected, but as the aftermath of Hurricane Katrina clearly demonstrated, issues such as who lives in areas that are safe and who does not, who can be evacuated and who cannot, who has resources to live on and who doesn't, who is immediately bailed out with social resources and who is left without schools, hospitals, or homes to live in . . . are all determined by class. It is the mode of production that determines what is "uncertain" and what is done in response to it. Market-driven social relations are the means by which "uncertainty" is perpetuated at the expense of the working class, while explanatory critique of such relationality is precisely what is dissipated by Harman's rendering of uncertainty as the basis of ontology. Despite the claim that he's discovered a new pathway to "third-way" thinking between revolution and reaction, Harman's mystical ontology dematerializes and dehistoricizes the world and ends up celebrating disorder and ignorance as the highest level of "freedom." It is not an accident, in other words, that Harman concludes a recent lecture by troping *The Communist Manifesto* in order to deny the call for working class unity, saying "I cannot close this talk by saying 'Objects of the world, unite!' because that is precisely what we do not want and cannot achieve."[75] What greater benefit to the ruling class could the proponents of the object-turn deliver than a cultural logic that in the guise of revolutionary thinking teaches its followers that nothing is connected, nothing can be known, and that there is no possible alternative to the existing since all existence is inherently alienated, divided, and without meaning?

The mystical celebration of the unknowability of objects is an instance of what Hegel describes in *The Science of Logic* as *understanding*, which rises above spontaneous intuition but only to arrive at a simplistic theory of the identity of objects. He writes, "thinking as *understanding* does not budge beyond the firm determinateness (of what is entertained) and its distinctness over against others" and "in the contemplation of nature, for instance, matters, forces, genera, etc. are distinguished and fixed as such in their isolation."[76] *Understanding* is a mode of thinking that focuses on isolation and nonrelation except in terms of difference, a conceptual framework celebrated by Harman when, for instance, he writes "to be an object does not mean to be physical material without dignity, but simply to be a unified entity irreducible to its component pieces or to its effects on the surrounding environment."[77] In contrast, Hegel shows why *understanding*, in its focus on identity and difference, cannot actually grasp the *dialectical* movement of contradictions in which "everything finite is this, the sublating of itself."[78] In other words, identity and difference exist not just between finite objects, but within them as contradictions as, for example, in the way in which a seed exists as a seed and yet also contains within it the contradictions that will negate its existence as a seed and mean it will either become a plant or lose its ability to become a plant. At the same time, whether a seed does or does not become a plant is not simply determined in itself, but in relations between the seed, soil, air, water, sunlight. . . . In other words, both identity and difference exist within the seed and without. It is only the dialectic, Hegel demonstrates, which can account for actual change by grasping the way in which "all finite things, instead of being something fixed and ultimate, are really changeable and perishable."[79] Harman says that things cannot be reduced to their essences, but it is the dialectic that grasps the logic of change.

It is Marx who first fully grasped the revolutionary potential of the dialectic. Demonstrating how *understanding* is the dominant ideology of existing relations, in which ideologues "actually [do] no more than interpret, systematise and defend in doctrinaire fashion the conceptions of the agents of bourgeois production who are entrapped in bourgeois production relations," it should not surprise us when thinkers, such as those promoting the object-turn, "[feel] particularly at home in the estranged outward appearances of economic relations in which these *prima facie* absurd and perfect contradictions appear and that these relations seem the more self-evident the more their internal relationships are concealed from it."[80] The issue is not that the connection between knowledge and reality are mechanical—which is the claim of the object-turn proponents who dismiss critique on the grounds that it "eliminates" chance—since "all science would be superfluous if the outward appearance and the essence of things directly coincided."[81] To approach the world *dialectically* means that we cannot assume that what exists has always existed or that what exists will never change, but rather we must grasp the cause of change itself if we are to "[replace] the domination of circumstances and of chance over individuals by the domination of individuals over chance and circumstances."[82] To assume a division between "objects" (human or nonhuman) as either eternal—as in the case of Harman's theory of universal alienation—or something that can be overcome through a more ethical approach between objects—as Bennett and others claim—is to begin with reality *post festum* and to attempt to deduce the future from its surface contradictions. Instead, we must begin with the existing and first uncover the root causes of the existing. It is necessary to examine the material conditions that constitute the world in which we live if we are to understand both the limits and the possibilities of the existing rather than accepting it as given.

Labor, Objectification and "Reality"

"Reality"—by which I mean the material structure of social, political, and economic relations that shape what is commonsensically represented as the "everyday"—is, as Marx writes, "nothing but the creation of man through human labor."[83] Regardless of whether it is food that is grown and harvested, oil that is removed from deep in the earth, software that is keeping track of the trillions of dollars exchanged on financial markets, or an abstract painting of a landscape, these objects would not exist as useful objects without human labor. Although "objectification" is often represented as a negative, in the sense of a moral reading that focuses on the consequences of capitalist production ("feeling objectified"), it is the ability of human labor to *objectify* the world that makes society possible. Objectification is the putting into practice of identity and difference through labor in order to transform the existing into something else. It is labor that turns trees into building materials, chemicals into life-saving medicines, and ideas into words and images which help us to expand our understanding of the world. What makes society possible, in short, is production; humans, who are part of nature, act on nature and the external material world in order to transform it to meet their needs (and in doing so produce new modes of production and change their needs). That the basis of "reality" is constituted by human labor means, I argue, that we cannot separate debates over what constitutes "reality" from the ways in which human labor is organized. This is what Marx referred to as the "social fabric" of humanity, namely the development of labor under determinate historical conditions.[84]

Under capitalism, it is the exploitation of labor that shapes all debates over the terms of reality, and yet this exploitation is obscured by the growth of the global market. The more that production goes global, the more that the products of labor are sent to markets far from their production, the more that technology is used to increase productivity while driving down wages, the more that the objects around us take on the appearance of "magic." We do not know where things come from or how they come into being, instead, they just appear before us as if nothing has produced them. Increasingly, those in the Global North live in an even more "ghostly" reality in which they do not even have to encounter others during acts of consumption. They shop on Amazon.com, using seemingly "immaterial" credit, and the commodity is delivered—possibly in the future by unmanned drones—all without necessarily ever having to see or talk with another person. This is what Marx refers to as the ideology of "commodity fetishism." It is the way in which the expansion of productivity under capitalism, which further drives workers away from both the products of their labor as well as from other workers, ideologically obscures the reality of exploitation and turns "a definite social relation between [people]" into "the fantastic form of a relation between things."[85] The alienated worldview reflected in the work of the object-turn, in which the products of labor come to confront workers "as an alien power,"[86] is the consequence of a society divided by exploitation.

Take, for example, Harman's reading of the "hammer." When Harman argues that "the hammer occupies a mezzanine level midway between the domestic relation of its pieces and the foreign relations of its outward effects" he concludes that "the hammer is not matter or effect . . . but a *substantial* form that is what it is quite apart from its current relations, and indeed apart from any relations at all."[87] For Harman, all objects are singular, autonomous, and ultimately isolated from all other objects. This, he claims, means that the hammer is "absolutely *not* knowable, thereby making it impossible to make any sweeping proclamations about the timeless essence or any individual or group."[88] Harman's solution to the instrumentalist reduction of the hammer is to reject *all* reduction, *all* relations, and *all* determination *tout court* and, instead, to assert that this is the essential means by which freedom and dignity for the object (and, by extension, the human-object) can be achieved. But, the problem is that a hammer is not an ahistorical object. It is an object created by labor in the past to be used by labor in the present. To declare, as Harman does, that its meaning is "nothing more than the tension between a real object and its real perception in a single instant"[89] is to simply side-step the question of why a hammer even exists, much less why it appears as an "alien" object. The fact that we can even talk about hammers and why they now appear as an object that is "alienated" from humanity is historical. As Marx explains,

> in the production process of capital, as will be seen more closely in its further development, labour is a totality—a combination of labours—whose individual components are alien to one another . . . The combination of this labour appears just as subservient to and led by an alien will and an alien intelligence—having its *animating unity* elsewhere—as its material unity appears subordinate to the *objective unity* of the *machinery*, of fixed capital, which, as *animated monster*, objectifies the scientific idea, and is in fact the coordinator, does not in any way relate to the individual worker as his instrument; but rather he himself exists as an animated individual punctuation mark, as its living isolated accessory.[90]

It is the structure of production under capitalism that separates labor from the products of labor through the private ownership of the means of production, before bringing labor together under the threat of economic compulsion owing to these same conditions of private ownership which mean the worker has nothing to sell on the market but their own labor-power. Under these conditions, in having to sell themselves for a wage, workers are alienated from themselves, from their fellow workers, as well as from the technology they must produce and employ. At every stage, it seems as if the individual, forced to compete in order to survive, is under the control of an alien power. It is not an accident, in other words, that Heidegger, Harman, and others who take up this line of thought tend to focus on the products of human labor—hammers, litter, poisoned rats, nuclear waste, electricity blackouts, or wider environmental destruction—while erasing the exploitative conditions under which the majority of humanity works. They take as their starting point the existence of the commodity and extend commodity culture to its logical conclusion by claiming that the displacement of the human-worker is somehow revolutionary thinking! This is the actual reality underlying the "ghostly" objects that seem to have a life of their own. The hammer is a product of labor whose uses are determined by the conditions of labor. Its appearances and disappearances from one person to the next are reflections of the market, not that it has a mysterious and unknowable "surplus" existence. It is in this context, for instance, that "instrumentality" has become a problem for capital today because a global market of production and consumption means that commodities need to live multiple lives—with the ability to be produced in one area and sold in another while accounting for local customs and market demands in both. Object-oriented philosophy provides the ideological "solution" to the problem of instrumental thinking by teaching readers to think of the object as not reducible to its positioning in local relations and instead as a site for the spontaneous generation of surplus cultural meanings. In this way, it enables precisely the kind of "out-of-the-box" thinking that is necessary to navigate global supply chains without questioning their exploitative purposes. In turn, the reduction of humanity to the commodity of labor power explains why we see in culture today the love/hate relation with the nonhuman object who seems equal and yet often has the appearance of more rights, more abilities, more power than workers today, whether it is the declaration that "corporations are people" or how "free-trade" agreements that allow for the free-flow of commodities across borders often come hand-in-hand with draconian restrictions on immigration and militarized police forces that begin with the assumption that working people are disposable objects. In contrast, as Marx writes, "man does not lose himself in his object . . . only when the object becomes for him a *social* object, he himself for himself a social being, just as society becomes a being for him in this object."[91] If today it seems as if the "strangeness" of the hammer has grown from a simply tool into massive "animated monsters" that exceed our understanding while threatening our very existence, it is a reflection of just how extensive the economic dictates of capital have come to dominate all aspects of reality.

Capitalism's "Animated Monsters"

Another of the key figures in the object-turn is Timothy Morton. Heavily influenced by Harman's work, the central thesis of Morton's *Hyperobjects: Philosophy and Ecology after the End of the World* is that humanity's instrumental drive to control the material world results in a paradox in which we are increasingly witness to "the weirdness of things, a weirdness that evolution, relativity, and quantum theory all speak out."[92] What, for Morton, makes the contemporary increasingly "weird" is that attempts to understand ontology from the quantum level to the scale of the universe have meant that "our reality has become more real, in the sense of more vivid and intense, and yet it has also become less knowable as some one-sided, facile thing."[93] Whether it is global warming, the (spooky) action at a distance of quantum mechanics, the digital algorithms which run the stock market, or non-biodegradable objects like plastic cups and spoons, he argues that we have "entered a new phase of history in which nonhumans are no longer excluded or merely decorative features of [humans'] social, psychic and philosophic space."[94] Instead, he proposes, objects are active and our growing awareness of nonhuman agency requires a new theory of ontology he describes as "*ecology without matter*."[95]

Leaving aside that the very awareness of this new "weird" reality Morton describes is, even on his own terms, the effect of human labor as manifest in science, at the core of Morton's argument for a nonmaterial ecology is the idea that an intensifying awareness of the coexistence of the human and the nonhuman is such that there is no longer any "outside" to the interrelations between the two from which the human can claim to know anything about the nonhuman. Taking as his starting point the division inaugurated by Kant between "beings of sense (*phaenomena*)"and "beings of understanding (*noumena*),"[96] Morton argues that we live in "an age of asymmetry," in which "the infinite powers of cognition and the infinite being of things" are forced to confront one another, leading to "a crazy arms race between *what we know* and *what is*, in which the technology of what we know is turned against itself."[97] In this, he not only repeats Kant's division between *phaenomena* and *noumena*, but, like Harman, also suggests that we must go even further in grasping the "radical unknowability" of the object. He proposes that an object is not simply a being of sense *or* of understanding, because both "sense" and "understanding" are versions of what he refers to as "correlationism" or "the notion that philosophy can only talk within a narrow bandwidth, restricted to the human-world correlate."[98] Instead, he argues, "objects compose an untotalizable nonwhole set that defies holism and reductionism," requiring that we "abolish the idea of the possibility of a metalanguage that could account for things while remaining uncontaminated by them."[99] In other words, there are no human-world, subject-object,

object-object relations that would necessitate the possibility of conceptual distance. Instead, in an echo of Heidegger's argument against instrumentality that "despite all conquests of distances the nearness of the thing remains absent,"[100] he states, "'distance is only a psychic and ideological construct designed to protect me from the nearness of things."[101] We are, in other words, surrounded by and enmeshed in networks of objects without any distance for conceptualization, and thus without the ability to know either the objects we confront as they confront us or the relations that lead to such confrontations.

What particularly concerns Morton is an increasing awareness of what he calls "hyperobjects" or "things that are massively distributed in time and space relative to humans."[102] What differentiates Morton from Harman is that in Morton's speculative framework there are still "relations" between objects, but he nonetheless declares that relations too are "objects" and therefore also withdraw from human understanding. Systems such as climate, the earth, the solar system, financial systems, black holes, capitalism . . . are considered just as autonomous and individual as the paper cup whose "being . . . is as profound" as any human.[103] This "flattening" of ontology, such that systems and objects operate on the same plane of existence while also withdrawing from one another, means that while Morton can make references to human agency as a disruptive cause of climate change, he nonetheless removes any historical understanding of the conditions producing environmental destruction. In the figure of global warming or financial collapse, for instance, Morton argues "two hundred years of idealism, two hundred years of seeing humans at the center of existence and now the objects take their revenge, terrifyingly huge, ancient, long-lived, threateningly minute, invading every cell in our body."[104] This appeal to "two hundred years of idealism" gives the illusion of a theory of causality while simultaneously erasing the fact that causality means making distinctions between the subjects of change and the objects of change, between material conditions and the ideas that people have about those conditions. Why just "two hundred years"? What does it mean that this period of "seeing humans at the center of existence" corresponds to capitalism's global ascendency? Instead of looking for the answers to such questions in causes that would explain both why such ideas took hold and why they are affecting the environment in damaging ways, he remains at the level of ideas and, claiming to look forward, proposes that we are witnessing the "end of the world." By this he does not mean an apocalyptic sense of environmental collapse, but rather the productive end of "a story we were telling ourselves . . . about how different groups were partitioned according to different horizons—concepts now revealed as ontic prejudices smuggled into the realm of ontology."[105] What Morton is saying here is that all "partitioning" or "hierarchies"—whether at the level of ideas or in social relations—are the effect of epistemology and lose their ability to divide once we embrace a flattened ontology in which all objects are declared equal to others in their isolation and unknowability. He concludes, "by embracing the hyperobjects that loom into our social space, and dropping Nature, *world*, and so on, we have a chance to create more democratic modes of coexistence between humans and with nonhumans,"[106] an ethical obligation based upon "the simple fact that existence is coexistence."[107] It is such that Morton's argument dehistoricizes the material conditions that make revolutionary coexistence possible, namely social transformation of the capitalist mode of production. This is because to presume to know the conditions of this new democracy of coexistence, following Morton's logic, is to presume the ability to determine cause and effect. At the same time, in reducing all ontology to a flat plane such that there is really no way to distinguish between the two, it means limiting this ability to the stochastic pragmatism of the local and the contingent rather than the global.

What Morton is proposing is essentially the same defense of "individualism" that one finds in humanism, not because posthumanism is merely humanism in disguise, but because both are the ideological expressions of capitalism. Under capitalism, "individualism" means nothing but the reduction of human relations to the "money relation."[108] While capitalism homogenizes individuals by turning them into interchangeable wage-laborers owing to the fact that the individual has nothing but the sale of their labor-power to survive, this reduced existence becomes "individualism" at the level of ideology precisely because capitalism turns individuals into alienated commodities that must compete with each other and not see in other individuals the collective agency of labor. Despite Morton's more "democratic" vision of "ethical co-existence" than what writers such as Harman or Brassier propose, he nonetheless reaches the same "weak" theory of social understanding as all of the other proponents of the object-turn. Writing about capitalism he states, "since the raw machinery of capitalism is reactive rather than proactive, it might contain a flaw that makes it unable to address the ecological emergency fully."[109] There are two problems with this statement. First, on an immanent level, if, as he says, "for every object in the universe there is a genuinely *future future* that is radically unknowable" and "a genuinely *elsewhere elsewhere* that is radically unknowable"[110] then the extent to which he can declare an "environmental catastrophe" at all much less diagnose its solutions would very much be in question. Morton, like others of the object-turn, continually relies on the labor of science to make prognostications even as he simultaneously dismisses, belittles, or erases the fundamental role of labor as scientific objectification in our understanding of the world. Second, however, is the more significant aspect of his weak resolution that capitalism *might* contain a flaw that makes it unable to address global warming. In fact, it is precisely his unknowing certitude that enables him to remain "open" to the possibility that the problem is not capitalism at all, but something else. Here is the true political "truth" of the object-turn. It is always "something else" for the North Atlantic left. They are always "open" to any possibility except the possibility that what is necessary is a revolution from the private ownership of capital to the collective ownership of workers. It is, for the proponents of the object-turn, "easier to imagine the end of the world than to imagine the end of capitalism."[111]

That capitalism is not opposed to the anti-instrumental and antihumanist theory of the object-turn is evident in just how much it mirrors contemporary popular culture in such films as 2014's *Godzilla*. The issue is not, of course, that Hollywood filmmakers are necessarily reading Heidegger, Bennett, Harman, and Morton (though the recently published emails between Morton and the singer Björk show that it is not impossible either).[112] Rather the issue is that both high theory and popular culture are reflections of the contradictions between the forces and relations of production and how these conflicts manifest in the class struggle at the level of ideas. What capitalism is confronting today is the very anarchic nature of production for profit which, as Marx and Engels write, "has conjured up such gigantic means of production and of exchange [and] is like the sorcerer, who is no longer able to control the powers of the nether world whom he has called up by his spells."[113] In this context, it is capitalism that is creating the object-turn by subjecting living labor to the demands of dead labor (in the form of technology and machinery that are used to drive up productivity while driving down wages).

The subjection of living labor to what Marx called "animated monsters"[114] is not just the literal plot of *Godzilla*, but its underlying logic. The film is a manifestation of a petty-bourgeois anxiety that modern capitalist society is creating meta-human conditions that, as Morton claims, result in the realization that "what we took to be a reliable world was just a habitual pattern."[115] Whether financial collapse, environmental changes, or rapid technological change, there is a fear that things are happening "too fast" and are "out of control" leading to the potential for the end of life as we know it (but with that end always represented as an apocalypse rather than as having the potential for radical change). During the opening credits the audience is told that that the particular kaijū ("monster") known as Godzilla is "a species of reptile that we thought was extinct"[116] but which appeared to be in some kind of hibernation until a 1954 experiment with a nuclear submarine accidentally awoke the creature. This event, an expression of what happens when humanity "goes too far," resulted in the United States' nuclear testing program that viewers are told was not part of the class war directed against the Soviet Union, China, and any other regimes which threatened the global imperialism of capital, but in actuality a secret attempt to kill the creature. Ultimately, the world appears reliably quiet until 2014 when another creature—referred throughout the film by the acronym "MUTO" (or "Massive Unidentified Terrestrial Organism")—awakens and begins searching out and feeding on nuclear power, with splitting the atom reemerging as the ultimate sign of the "hubris" of humanity. When a second MUTO awakens, it becomes clear that they are going to meet in order to reproduce, creating not only a trail of destruction as they travel to join one another but also the potential for massive future devastation and a posthuman world. In response, the U.S. military devises a plan for dealing with the nuclear-eating animals that requires the launch of a nuclear weapon at them. This absurd expression of instrumentality—using the creature's food source in an attempt to kill them—ultimately fails, of course, and it is up to the nonhuman hero of the film, Godzilla, to "save" humanity from itself. After defeating and killing the two MUTOs, Godzilla eventually heads back to the ocean and disappears once again while the audience sees a news clip in the background asking whether Godzilla is "King of the Monsters: Savior of the City?"

While the film's narrative includes an overly sentimental and wildly coincidental human subplot as a formal means of getting the audience to identify with the events of the film, what is significant about the "human" aspects of the story is just how powerless and ignorant humanity is about these new conditions of existence. The reasons why the MUTOs emerge when they do and why Godzilla also arises to fight and kill them is never fully explained but is given a weak theory that Godzilla is part of a "natural order" and exists to "restore balance." The film couches the events in terms of the spiritual—Godzilla as "a god, for all intents and purposes" as one character states[117]—because we are meant to conclude that things will "work out" but not because of anything in the power of humanity. Instead, humanity's only response to the "hyperobjects" fighting one another is the logic of instrumentality—to attempt to use nuclear weapons (as a sign for the willingness on the part of humanity to end all life if it threatens human life) to combat a nonhuman that cannot be contained by human science or technology. The character of Dr. Ichiro Serizawa sums up the film's conclusions when he declares, to the consternation of the military leaders, "the arrogance of men is thinking nature is in our control and not the other way around. Let them fight."[118]

Humanity "wins" in the film, just as Morton declares, when "we see that we are *weak*, in the precise sense that our discourse and maps and plans regarding things are not those things."[119] In other words, that we are given no explanation for the events of the film and that humanity seems at such a loss to understand or address any of these issues except futility is not a plot oversight. It is a reflection of the object-turn's idea that "the more data we have about lifeforms, the more we realize we can *never* truly know them," as Morton proposes.[120] Morton's theory of "hyperobjects" and the film's depiction of out-of-control, nonhuman agents are expressions of a global world that is being forced to confront massive problems—financial, environmental, military, economic, cultural—that exceed borders and boundaries, but with no way of grasping these problems as elements in a dialectical totality. For the proponents of the object-turn, like the characters in the film, the new ontology of globalization means "the end of the world [is] the beginning of an uncertain, hesitating futurality"[121] that requires a "radical letting go of what constitutes a self" because "I can never fully experience, explain, or otherwise account for her, him, or it."[122] In other words, "unknowing" action is the very expression of the (so-called) "end of the world" that will require a new mode of "affective coexistence." Both the film and Morton envision a "new" reality that all prior epistemologies fall short in grasping. Whether it is Godzilla or financial markets, we are told that humanity is creating monsters that it can no longer control, understand, or even challenge without just making things worse. To survive, in short, ultimately means to give up believing that any sort of human agency—regardless of the ends to which it is put—can address the contradictions emerging as a result of the nonhuman actants operating outside of the limits of human relations.

What links all of these arguments is the notion that central to the contradictions of contemporary capitalism is a *cultural* intentionality that, on the one hand, subjects the nonhuman to a destructive instrumental logic and, on the other, leaves humanity at the mercy of the same nonhuman "hyperobjects" that human instrumentality created but can no longer know, control, or transform. In response, the logic of the object-turn posits the need to embrace locality, singularity, and an inability to totalize as the means by which to challenge the instrumental logic of capitalism. Just as the film turns the deadly threat of the use of nuclear weaponry on the world's populations by the United States into an undecidable epistemology that *tries* to "do good" but ultimately backfires, not because it is part of the state apparatus used to maintain the control of capital over labor, but because it fails to capture the natural "balance" of noninstrumental thinking, the object-turn displaces capitalism's contradictions by ontologizing the contradictions and uncertainty of life under capitalism as the basis of life "as such." "Instrumentality," however, is not the purpose of capitalism and is not simply a logical flaw that eliminates the agency of the nonhuman object. It is a managerial logic that emerges as an effect of the organization of production for profit and can be modified if new forms are necessary for profit production, as we are witnessing today. In fact, the change from instrumental to noninstrumental managing is what is currently being promoted in books like *Work Rules!* and *Holacracy* as moving away from the "command and control" of industrial capitalism to "a new way of running an organization that removes power from a management hierarchy and distributes it across clear roles, which can then be executed autonomously, without a micromanaging boss."[123] The object-turn is the testing ground for the new management strategies necessary for a global market.

The growing concern with the nonhuman is a response to the alienation of labor under capitalism, but it does not provide an alternative to it. Far from challenging these contradictions, the underlying logic of "ethical co-existence" between the human and the nonhuman naturalizes the conditions of alienation and becomes a way of "re-animating" the mysteriousness of the commodity. If capitalism is "animating monsters" that seem to be operating out of control, the solution is not the "ghostly objectivity" of the object-turn that denies the ability or know or understand and places all objects beyond the reach of the critique of social relations. It is time, as Marx says, "to let the dead bury their dead."[124] It is only by understanding the contradictions of capitalism in relation to their underlying logic in exploitation that objects will no longer appear as having a mystical and independent existence against and above humanity, the majority will be able to change the conditions of the world, and society will be directed toward meeting the needs of all.

BIBLIOGRAPHY

Atkin, Emily. "'I'm Not A Scientist': A Complete Guide To Politicians Who Plead Ignorance On Climate Change." *Thinkprogress.org*. October 3, 2014. http://thinkprogress.org/climate/2014/10/03/3575849/not-a-scientist/.

Beck, Ulrich, and Elisabeth Beck-Gernsheim. *Individualization: Institutionalized Individualism and Its Social and Political Consequences*. London: SAGE, 2002.

Bennett, Jane. *Vibrant Matter: A Political Ecology of Things*. Durham, NC: Duke University Press, 2010.

Bock, Laszlo. *Work Rules!: Insights from Inside Google That Will Transform How You Live and Lead*. New York: Twelve Books, 2015.

Bogost, Ian. *Alien Phenomenology; or, What It's Like to Be a Thing*. Minneapolis: University of Minnesota Press, 2012.

Brassier, Ray. *Nihil Unbound: Enlightenment and Extinction*. Basingstoke, UK and New York: Palgrave Macmillan, 2010.

Brown, Bill. "Thing Theory." *Critical Inquiry* 28, no. 1 (2001): 1–22.

Bryant, Levi R. *The Democracy of Objects*. Ann Arbor, MI: Open Humanities Press, 2011.

Candlin, Fiona, and Raiford Guins. "Introducing objects." In *The Object Reader*, edited by Fiona Candlin and Raiford Guins, 1–18. London and New York: Routledge, 2009.

Cohen, Jeffrey J. *Stone: An Ecology of the Inhuman*. Minneapolis: University of Minnesota Press, 2015.

Connolly, William E. *The Fragility of Things: Self-organizing Processes, Neoliberal Fantasies, and Democratic Activism*. Durham, NC: Duke University Press, 2013.

DeLanda, Manuel. *Intensive Science and Virtual Philosophy*. New York: Bloomsbury, 2013.

Derrida, Jacques. *Margins of Philosophy*. Translated by Alan Bass. Chicago, IL: University of Chicago Press, 1982.

---. *Of Spirit: Heidegger and the Question*. Translated by Geoffrey Bennington and Rachel Bowlby. Chicago, IL: University of Chicago Press, 1991.

Garcia, Tristan. *Form and Object: A Treatise on Things*. Translated by Mark A. Ohm, and Jon Cogburn. Edinburgh, UK: Edinburgh University Press, 2014.

Godzilla. Directed by Gareth Edwards. 2014; Hollywood, CA: Warner Bros. Home Video, 2014. Blu-Ray.

Harman, Graham. *Bells and Whistles: More Speculative Realism*. Winchester, UK: Zero Books, 2013.

---. *The Quadruple Object*. Winchester, UK: Zero Books, 2011.
Hegel, Georg W. F. *Encyclopedia of the Philosophical Sciences in Basic Outline: Part 1.* Translated and Edited by Klaus Brinkmann and Daniel O. Dahlstrom. Cambridge: Cambridge University Press, 2010.
Heidegger, Martin. *Being and Time*. Translated by John Macquarrie and Edward Robinson. Oxford, UK and Cambridge, MA: Blackwell Publishers, 2001.
---. "The Question Concerning Technology." In *Basic Writings*, translated by David Farrell Krell, 307–42. San Francisco, CA: Harper, 1993.
---. "The Thing." In *Poetry, Language, Thought*, translated by Albert Hofstadter, 161–84. New York: HarperCollins, 2001.
Holacracy.org. "How it Works," *Holacracy*. Accessed July 19, 2015. http://www.holacracy.org/how-it-works/.
Jameson, Frederic. "Future City." *New Left Review* 21 (2003): 65–79.
Lukács, György. *History and Class Consciousness: Studies in Marxist Dialectics*. Cambridge, MA: MIT Press, 1971.
Marx, Karl. *Capital Volume 1.* In *Marx-Engels Collected Works*, vol. 35. New York: International Publishers, 1996.
---. *Capital Volume 3.* In *Marx-Engels Collected Works*, vol. 37. New York: International Publishers, 1998.
---. *Economic and Philosophic Manuscripts of 1844*. In *Marx-Engels Collected Works*, vol. 3, 229–346. New York: International Publishers, 1975.
---. *The Eighteenth Brumaire of Louis Bonaparte*. In *Marx-Engels Collected Works*, vol. 11, 99–197. New York: International Publishers, 1979.
---. *Grundrisse.* Translated by Martin Nicolaus. New York: Vintage Books, 1973.
Marx, Karl, and Frederick Engels. *The German Ideology*. In *Marx-Engels Collected Works*, vol. 5, 19–584. New York: International Publishers, 1976.
---. *Manifesto of the Communist Party*. In *Marx-Engels Collected Works*, vol. 6, 477–519. New York: International Publishers, 1976.
Meillassoux, Quentin. *After Finitude: An Essay on the Necessity of Contingency*. Translated by Ray Brassier. London: Continuum, 2009.
Morton, Timothy. *Hyperobjects: Philosophy and Ecology after the End of the World*. Minneapolis: University of Minnesota Press, 2013.
Robertson, Brian J. *Holacracy: The New Management System for a Rapidly Changing World*. New York: Henry Holt and Company, 2015.
Sax, Boria. *Imaginary Animals: The Monstrous, the Wondrous and the Human*. London: Reaktion Books, 2013.
Shaviro, Steven. *The Universe of Things: On Speculative Realism*. Minneapolis: University of Minnesota Press, 2014.
Simondon, Gilbert. *Two Lessons on Animal and Man*. Trans. Drew S. Burk. Minneapolis: University of Minnesota Press, 2012.

NOTES

1. Karl Marx, *The Eighteenth Brumaire of Louis Bonaparte*, in *Marx-Engels Collected Works*, vol. 11 (New York: International Publishers, 1979), 104.
2. Genesis 2:7.
3. Ulrich Beck and Elisabeth Beck-Gernsheim, *Individualization: Institutionalized Individualism and Its Social and Political Consequences* (London: SAGE, 2002), 203.
4. Jane Bennett, *Vibrant Matter: A Political Ecology of Things* (Durham, NC: Duke University Press, 2010), ix.
5. Ibid., ix.
6. Ibid., 5.
7. William E. Connolly, *The Fragility of Things: Self-organizing Processes, Neoliberal Fantasies, and Democratic Activism* (Durham, NC: Duke University Press, 2013), 3.

8. Jacques Derrida, *Margins of Philosophy*, trans. Alan Bass (Chicago, IL: University of Chicago Press, 1982), 1–27.
9. Bennett, *Vibrant Matter*, x.
10. Ibid., viii.
11. Connolly, *The Fragility of Things*, 9.
12. Karl Marx, *Economic and Philosophic Manuscripts of 1844*, in *Marx-Engels Collected Works*, vol. 3 (New York: International Publishers, 1975), 300.
13. Karl Marx, *Grundrisse*, trans. Martin Nicolaus (New York: Vintage Books, 1973), 489.
14. Jeffrey J. Cohen, *Stone: An Ecology of the Inhuman* (Minneapolis: University of Minnesota Press, 2015), 8.
15. Marx, *Grundrisse*, 461.
16. Bennett, *Vibrant Matter*, 20.
17. Steven Shaviro, *The Universe of Things: On Speculative Realism* (Minneapolis: University of Minnesota Press, 2014), 12.
18. Ibid., 67.
19. György Lukács, *History and Class Consciousness: Studies in Marxist Dialectics* (Cambridge, MA: MIT Press, 1971), 83.
20. Graham Harman, *Bells and Whistles: More Speculative Realism* (Winchester, UK: Zero Books, 2013), 39.
21. Karl Marx, *Capital Volume 1*, in *Marx-Engels Collected Works*, vol. 35 (New York: International Publishers, 1996), 321.
22. Harman, *Bells and Whistles*, 16.
23. Fiona Candlin and Raiford Guins, "Introducing objects," in *The Object Reader*, ed. Fiona Candlin and Raiford Guins (London and New York: Routledge, 2009), 2.
24. Ian Bogost, *Alien Phenomenology or What It's Like to Be a Thing* (Minneapolis: University of Minnesota Press, 2012), 109.
25. Bennett, *Vibrant Matter*, xiv.
26. Bill Brown, "Thing Theory," *Critical Inquiry* 28, no. 1 (2001): 5.
27. Tristan Garcia, *Form and Object: A Treatise on Things*, trans. Mark A. Ohm and Jon Cogburn (Edinburgh, UK: Edinburgh University Press, 2014), 7.
28. Manuel DeLanda, *Intensive Science and Virtual Philosophy* (New York: Bloomsbury, 2013), 51.
29. Bogost, *Alien Phenomenology*, 133.
30. Levi R. Bryant, Levi R, *The Democracy of Objects* (Ann Arbor, MI: Open Humanities Press, 2011), 19.
31. Bennett, *Vibrant Matter*, ix.
32. Ibid., 108.
33. Martin Heidegger, "The Question Concerning Technology," in *Basic Writings*, trans. David Farrell Krell (San Francisco, CA: Harper, 1993), 322.
34. Martin Heidegger, "The Thing," in *Poetry, Language, Thought*, trans. Albert Hofstadter (New York: HarperCollins, 2001), 161.
35. Ibid., 161.
36. Ibid., 168.
37. Martin Heidegger, *Being and Time*, trans. John Macquarrie and Edward Robinson (Oxford and Cambridge, MA: Blackwell Publishers, 2001), 96.
38. Heidegger, "The Thing," 179.
39. Ibid., 179.
40. Ibid., 166.
41. Ibid., 167.
42. Ibid., 177.
43. Heidegger, *Being and Time*, 98.
44. Ibid., 98.
45. Ibid., 102.
46. Ibid., 102
47. Ibid., 105.

48. Cohen, *Stone*, 4.
49. Boria Sax, *Imaginary Animals: The Monstrous, the Wondrous and the Human* (London: Reaktion Books, 2013), 32.
50. Shaviro, *The Universe of Things*, 105.
51. Ray Brassier, *Nihil Unbound: Enlightenment and Extinction* (Basingstoke, UK and New York: Palgrave Macmillan, 2010), xi.
52. Bennett, *Vibrant Matter*, xiv.
53. Harman, *Bells and Whistles*, 35.
54. Ibid., 5.
55. Ibid., 33.
56. Ibid., 61.
57. Ibid., 32.
58. Ibid., 126.
59. Graham Harman, *The Quadruple Object* (Winchester, UK: Zero Books, 2011), 52.
60. Ibid., 62.
61. Ibid., 73–74.
62. Ibid., 47.
63. Ibid., 44.
64. Ibid., 39.
65. Ibid., 44.
66. Ibid., 73.
67. Ibid., 73.
68. Harman, *Bells and Whistles*, 101.
69. Ibid., 124–25.
70. Ibid., 116.
71. Ibid., 125.
72. Emily Atkin, "'I'm Not A Scientist': A Complete Guide To Politicians Who Plead Ignorance On Climate Change," *Thinkprogress.org*, October 3, 2014, http://thinkprogress.org/climate/2014/10/03/3575849/not-a-scientist/.
73. Harman, *Bells and Whistles*, 120.
74. Ibid., 117.
75. Ibid., 39.
76. Georg W. F. Hegel, *Encyclopedia of the Philosophical Sciences in Basic Outline: Part 1*, trans. and ed. Klaus Brinkmann and Daniel O. Dahlstrom (Cambridge: Cambridge University Press, 2010), 126.
77. Harman, *Bells and Whistles*, 39.
78. Hegel, *Encyclopedia*, 129.
79. Ibid., 130.
80. Karl Marx, *Capital Volume 3*, in *Marx-Engels Collected Works*, vol. 37 (New York: International Publishers, 1998), 804.
81. Ibid., 804.
82. Karl Marx and Frederick Engels, *The German Ideology*, in *Marx-Engels Collected Works*, vol. 5 (New York: International Publishers, 1976), 438.
83. Marx, *Economic Manuscripts*, 305.
84. Ibid., 298.
85. Marx, *Capital Volume 1*, 321.
86. Marx, *Economic Manuscripts*, 278.
87. Harman, *Bells and Whistles*, 21.
88. Ibid., 27.
89. Ibid., 27.
90. Marx, *Grundrisse*, 470.
91. Marx, *Economic Manuscripts*, 301.
92. Timothy Morton, *Hyperobjects: Philosophy and Ecology after the End of the World* (Minneapolis: University of Minnesota Press, 2013), 159.
93. Ibid., 130.
94. Ibid., 22.

95. Ibid., 92.
96. Immanuel Kant, *Critique of Pure Reason*, trans. Paul Guyer and Allen W. Wood (Cambridge: Cambridge University Press, 1998), 306.
97. Morton, *Hyperobjects*, 22.
98. Ibid., 9.
99. Ibid., 2.
100. Heidegger, "The Thing," 164.
101. Morton, *Hyperobjects*, 27.
102. Ibid., 1.
103. Ibid., 17.
104. Ibid., 115.
105. Ibid., 103.
106. Ibid., 121.
107. Ibid., 125.
108. Karl Marx and Frederick Engels, *Manifesto of the Communist Party*, in *Marx-Engels Collected Works*, vol. 6 (New York: International Publishers, 1976), 487.
109. Morton, *Hyperobjects*, 21.
110. Ibid., 61.
111. Frederic Jameson, "Future City," *New Left Review* 21 (2003): 76.
112. Bjӧrk, Klaus Biesenbach, Alex Ross, Nicola Dibben, and Sjón, *Bjӧrk* (New York: The Museum of Modern Art, 2015).
113. Marx and Engels, *Manifesto*, 489.
114. Marx, *Grundrisse*, 470.
115. Morton, *Hyperobjects*, 102.
116. *Godzilla*, directed by Gareth Edwards (2014; Hollywood, CA: Warner Bros. Home Video, 2014), blu-ray.
117. Ibid.
118. Ibid.
119. Morton, *Hyperobjects*, 133.
120. . Ibid., 56.
121. Ibid., 95.
122. Ibid., 123.
123. Holacracy.org, "How it Works," accessed July 19, 2015, http://www.holacracy.org/how-it-works/.
124. Marx, *The Eighteenth Brumaire*, 106.

III

Theory in the Common, Theory in the Commune

SIX

The Commune, NOT the Common

Kimberly DeFazio

The posthumanist narrative dominant in the North Atlantic metropole today posits the continuation of the human species as contingent upon a series of disasters affecting the basis of life in common: climate change, biogenetic change, intellectual property, and global inequality. Influenced by "common-ist" writers such as Antonio Negri and Michael Hardt (*Empire*; *Commonwealth*), Alain Badiou (*The Communist Hypothesis*), Slavoj Žižek (*In Defense of Lost Causes*), and Jodi Dean (*The Communist Horizon*), the "common" has emerged on the transatlantic left as a way to make sense of the diverse crises that capitalism is producing but is unable to solve. In place of connecting contemporary crises to their underlying material causes, the commonist left champions "common" modes of life that, it is claimed, represent an "exodus" from capitalist relations. This is another way of saying that left commonist theory, in a mirror-image of the neoliberal declaration that "there is no alternative," rejects the struggle to transform capitalism and suggests instead that the only option for those who seek to resist the heightening contradictions of capitalism is "exiting" or "subtraction" from class society into the immateriality of a fluid and indeterminate conception of life itself that accommodates living with others. Essentially a spiritualist reaction to material contradictions, commonist theory in effect advocates retreat into "castles in the air" rather than realize the revolutionary potential of labor on earth.[1]

The common in dominant left theory is associated with resisting the privatization of the products of nature and culture (e.g., air, land, water, as well as knowledges, languages, information, affects), and negating the neoliberal market logic of calculation that reduces all life to the same. In other words, the common has become the sign for the affirmation of the ethical project of sharing carried out by the multitude, which, in the words of Hardt and Negri, "makes itself by composing in the common" its "singular subjectivities" and which engages in "practices of interaction, care, and cohabitation in a common world."[2] According to commonist theory, humanity has reached the stage of production where its continuance as a species requires recognizing the biopolitics of its own being (the creativity of life itself) and the ways in which the human has always been defined against its "others" (e.g., animals, machines, the inhuman, and the nonhuman). What is necessary, as Stephen Healy puts it, is "[a]dmitting to the inescapability of our being-in-common" and "abandoning a homocentric conception of planetary well-being and learning to live in common with the biotic and abiotic forces that create conditions felicitous to life."[3] The primary concern is not overthrowing capitalism as a mode of production, but affirming the sites within capitalism that, by producing being-in-common, purportedly negate its homogenizing and privatizing forces.

For example, in their reading of J. M. Coetzee's *The Life and Times of Michael K,* a romantic anti-urban narrative based upon the philosophy that "a man must live so that he leaves no trace of his living,"[4] Hardt and Negri represent the central character of Michael K as "a simple man" that "appears to be not of this world."[5] They find that his desire to "keep moving [. . .] in order to keep life in motion" against the backdrop of "cages, barriers, and checkpoints erected by authority," is "the beginning of a liberatory politics."[6] His desire to retreat from modern life is, for them, emblematic of the figure of the multitude, a "fluid and amorphous" post-class grouping which is "shot through with lines of conflict and antagonism, but none that appear as fixed and eternal boundaries."[7] Against the forces of empire, the "decentered and deterritorializing apparatus of rule that progressively incorporates the entire global realm within its expanding frontiers,"[8] Hardt and Negri see in Michael K's refusals to submit to authority a prophet of a new vitalism which will "construct a new mode of life and above all a new community."[9] In their writing, going off the grid (a consumerist solution to capitalist crisis that has itself spawned entire new industries) re-emerges as a form of radical activist biopower.

Hardt and Negri's call for a "real alternative" to existing social relations that is "enriched by the collective intelligence and love of the community"[10] is, like other readings of the "community-to-come," premised upon the existence of "common" assemblages, practices, and spaces which, it is claimed, resist the daily mechanisms of control that restrict and divide up the flows of contemporary life. In commonist theory, social inequality is a product of how disciplinary power has essentialized a master narrative about what it means to be human that has served to deny the humanity of other humans as well as the significance of nonhuman life forms and their treatment as "things" to be used. The solution to global inequality is, on these terms, to abolish humanist ("homocentric") distinctions by realizing that with the advent of a global communications culture the means of social production are (already) held in common and are being used to produce a more fluid lifeworld in which all past distinctions, while not finally abolished, have become open to performative reversals, exodus, and new assemblages of affinities. These commonist assumptions about inequality are not limited to the writings of Hardt and Negri, however. In fact, so common is common theory today that it has almost become impossible to talk about inequality as other than inequality of "access" to the cultural common, as inclusion in or exclusion from the circuits of immaterial labor. From autonomist Marxist theorists like Hardt and Negri, who focus on the cultural common, to free market libertarians like Jimmy Wales (founder of Wikipedia), and the 2009 winner of the Nobel prize for economics for her work on the commons, Elinor Ostrom (who is herself largely inspired by Friedrich Hayek)[11]—commonism has become the dogma of the posthumanities on inequality.

In this essay I argue that commonist theory dissolves class consciousness in a biopolitical notion of community, valorizing local acts of commoning within capitalism that actually forestall social transformation. Instead, I propose an alternative reading of the "common" that moves the term in another class direction, away from the accommodationism of the common as "community" and toward the common as communism (as in the Paris Commune). The commune, rooted in the collectivity of the working class, is as Marx writes, "a lever for uprooting the economical foundation upon which rests the existence of classes, and therefore of class rule."[12] What makes the commune revolutionary is not an immaterial desire for exodus and escape, but the material collectivity of labor that takes hold of the means of production and, in the process, becomes the means by which to bring about the end of all forms of class exploitation.

To put my critique another way, by submerging class in a posthumanist notion of community, common theory's representation of capitalism as a regime of exclusion from the common obscures the root relations of inequality. Inequality is an effect of the deepening division of the world into those who own the means of production (factories, land, natural resources, technologies), thereby compelling others to work for them to create profit, and those who, owning nothing but their labor power, are forced to work for the owners to survive. Common theory cuts off inequality from the exploitation of wage labor—from the social totality of capitalism—and focuses instead on the ("common") relations of consumption that are viewed as undermining capitalist accumulation. Such arguments are popular because they make it impossible to understand the connection between deepening exploitation, deepening divisions along the lines of race, gender, sexuality, ability, religion, and age, for instance, and more and more catastrophic environmental crisis. The logic of inclusion and exclusion from the common does not explain but merely re-describes—with an ethical tone that puts shared feelings before critique-al concepts in cultural theory—inequality, which is the result of class. It is not lack of access to healthcare, for instance, that explains why millions of people die each year from treatable illnesses, while others enjoy boutique medicine; focusing on lack of access to or exclusion from healthcare begins with (unequal) distribution of products of consumption and ignores the exploitative relations within which products are produced. Inequalities in healthcare are the outcome of social arrangements in which owners of capital, in their perpetual quest to increase the rate of exploitation, transform the world's workers into (property-less) wage laborers, and turn ever more products of social need into exchange value. Starting with (unequal) consumption leads to the utopian conclusion that healthcare can be commonly shared by all members of society without abolishing wage labor, which is what determines not only access to healthcare but also the physical and mental stress that leads to the diseases affecting public health in the first place.

While much of the posthumanist turn to the common often draws from or aligns itself with the writings of Marx, as I discuss at further length below, it is worthwhile pointing out here that the "common" in Marx is not only a mode of co-habitation or state of being (which is a concern with culture and the relations of reproduction). It is, above all, the result of cooperation of labor at the point of production. Although he discusses the emergence of "communist consciousness" as a result of the dominance of the world market and global capital accumulation in *The German Ideology*,[13] for Marx the communist consciousness that results when the world market liberates individuals from their "various national and local barriers" by bringing them "into practical connection with the production (including intellectual production) of the whole world," is the result of the "*world-historical* co-operation" in production.[14] Capitalist property relations bring workers together for the primary purpose of appropriating their surplus value, the dialectical result of which is that the more global production becomes, leading to more "common" forms of culture (from the serialized novel and the diorama to film, Facebook, and periscoping), the greater the exploitation of the proletariat and the deeper they sink into poverty, the more the accumulation of social wealth grows in the hands of the few. Through this process is produced "the consciousness of the necessity of a fundamental revolution, the communist consciousness" that arises "through the contemplation of the situation of this class [. . .] the majority of all members of society [. . .] which has to bear all the burdens of this society without enjoying its advantages."[15] On this basis, Marx also reads the Paris Commune as a "glorious harbinger of a new society"[16] because it demonstrates that "*it was only the working class that could* formulate by the word 'Commune' and initiate by the fighting Commune of Paris" the emancipation of the exploited population.[17] The contradictions of the common under capital are in this way inextricably linked to the dialectic of capital and labor. Whereas commonist readings of the Paris Commune either reject the revolutionary working class organization of the Commune or treat it primarily in the cultural terms of biopower and affect that serve to glorify "spontaneous communism" within capitalism,[18] the commune represents in Marx the urgency of dialectical understanding of the wage labor-capital relation as a guide in the struggle for the collective ownership of property.

Cutting off common forms of culture from class so as to celebrate "spontaneous communism," commonists like Hardt and Negri suggest that, unlike the working class collectivity which established the Paris Commune, the fight over the common reflects the singularities and desires of the multitude, a fluid series of contingent groupings "in which there is no center, only an irreducible plurality of nodes in communication with each other."[19] The contingency of the multitude means that future societies cannot assume the Paris Commune's model of "the sovereignty of the people" since it inevitably leads to "the creation of political vanguards."[20] The task of theory, on these terms, is to affirm the biopower and spontaneous acts of resistance of the multitude. In effect, from this standpoint, it is not the commune but the common that makes Paris in 1871 significant: that is, how the "flesh" of the multitude and what Rimbaud called its "swarm" intelligence models more emancipatory networks of resistance to capital's capturing of the common.[21]

The common, and commonist readings of the commune, are at the center of the increasingly influential "new communist" theory. Broadly speaking, new communist theory includes such writers as Badiou, Dean and Žižek; Negri and Hardt and other post-*operaismo* and autonomist Marxists; "communisation" theory, as articulated in such journals as *Endnotes* and *SIC*; and the more recent "accelerationists" (e.g., Alex Williams, Nick Srniek, Ray Brassier) among others. In their writings, the common emerges as the basis of a new form of communism immanent within capitalism and thus as the ground of a revolutionary politics. But, as Lenin argues, "without revolutionary theory there can be no revolutionary *movement*,"[22] and the deeper class significance of the new communism, I argue, is that it represents the hijacking of revolutionary theory by a posthumanist theory of the common that is itself founded upon the primary assumptions of cognitive capitalism that privilege the petty bourgeois knowledge workers in contemporary capitalism and their new forms of consumption. Common theory is the latest apologetics of capitalism masquerading as radical theory. Its solution to global inequality ultimately puts forward the consumer branding of identity as a purported self-creative commons whose line of flight "escapes" the negative dialectic of class.

The Common and Cognitive Capitalism

In general terms, discussions of the common today tend to take either a more romantic approach to the natural common (often the natural "commons" of water or air, for instance) or a cultural approach to the common that addresses the common produced within capitalism: for instance, open access software, new forms of digital culture, and the fluid, decentered identities that have developed through social media. The most influential left theories of the common in the tradition of neo-, post-, and autonomist Marxism—the "new communists"—focus on the ways capital mediates humans' relation to nature, making the common not a natural essence prior to the social but one produced in and through the social. Thus Negri remarks, "We all know that the common is water, air, soil, the sea, land, etc. but we are interested in the human-made common, because the naturalist common has long been reabsorbed in the productive structures of capitalism."[23] His concern, like that of other autonomist Marxists, is "the common inside the multitude," where "the common does not precede or follow the multitude: the making of the multitude *is* the common."[24] While there are of course immanent debates within this context,[25] I argue that all ultimately posit the common as rooted in the cooperative *knowledges* and culture developed through technology, appealing to an anti-essentialist view of subjectivity produced in the common as a resistant, excessive and creative life force, assumedly "autonomous" from deeper-lying relations of production. The agent of the common, on these terms, is a posthumanist subject of biopolitics: a producer of creative, resistant life, a subject for whom the common is a post-class space of community resisting the *dispositifs* of power.

New communist theory takes as its starting point the presupposition that class exploitation under capitalism has been superseded by what Paolo Virno calls "the communism of capital."[26] This is based on the argument that capitalism is no longer organized around wage labor relations but power, and value no longer derives from material labor but knowledge, affect and communication (that is, "immaterial labor"). The exploitative property relations of capital, Virno suggests, have been replaced with a command economy of automation that eliminates wage labor, dematerializes the state, and incorporates all prior forms of resistance within it by turning the source of value from the extraction of surplus value into the control of surplus desires through the multiplication of subjectivities. It thereby "achieves" in negative form what past forms of communism had set out to accomplish. In other words, capital's tendencies toward monopoly concentration are interpreted as organizing "for its own benefit precisely those material and cultural conditions which would guarantee a calm version of realism for the potential communist."[27] The "new communists" seek to distance themselves from the "communism of capital" and to instead focus on the forces of creative labor that they argue are the basis of a truly communist society.

In its claim that knowledge is now the source of value today, the new communists repeat the mantras of cognitive capitalism, which, like its earlier iterations in writings of the proponents of "the knowledge economy" and "the network society," is represented as a radically new "third type of capitalism"[28] rooted in knowledge production: a "new historical and contemporary capitalism" "forged in Silicon Valley."[29] According to Yann Moulier Boutang, who is associated with autonomist Marxism, with the development of computer and information technologies in the 1970s, capitalism has mutated from an economy based on industrial production to one in which, as he puts it, the "object of accumulation consists mainly of knowledge, which becomes the basic source of value."[30]

Such claims about the role of "immaterial labor" have of course been central to the early work of Negri, who more recently with Hardt has been extremely influential in developing this theory into a notion of the common as the effect of knowledge produced through digital technologies. As they put it in *Empire*,

> today we participate in a more radical and profound commonality than has ever been experienced in the history of capitalism. The fact is that we participate in a productive world made up of communication and social networks, interactive services, and common languages. Our economic and social reality is defined less by the material objects that are made and consumed than by co-produced services and relationships. Producing increasingly means constructing cooperative and communicative commonalities.[31]

This is because, in Maurizio Lazzarato's words, "Cooperation between minds, unlike cooperation in the Smithian and Marxian factory, produces public, collective or common goods: knowledge, language, science, culture, art, information, forms of life, relations with oneself, others and the world etc."[32] On these terms, to accumulate value, capital less and less exploits labor at the point of production; more and more, capital is forced to "capture" value that is created in the common that is produced outside of productive labor, not only in immaterial labor but particularly in the creative processes "outside" of work that involve social media and other kinds of collaborations. This is part of a broader reading which sees production of value taking place in society more generally: "work and life are no longer separate; society is collapsed into the logic and processes of capitalist development," Negri and Félix Guattari argue in *Communists Like Us*.[33] "Society," they explain, increasingly becomes "a giant factory,"[34] in which what is produced now is "subjectivity." In the era of cognitive capitalism, Hardt and Negri write, the "core of biopolitical production [. . .] is not the production of objects for subjects, as commodity production is often understood, but the production of subjectivity itself."[35] The shift from "material" to "immaterial" production, as a result, involves biopolitical production, since "life is what infuses and dominates all production. In fact, the value of labor and production is determined deep in the viscera of life."[36]

At stake here is the question of "value": value rather than being a social relation based on class antagonism (exploitation) becomes an immaterial "thing" in itself that emerges spontaneously from the "bodies" of the (classless) "community." There is thus an economic argument that posthumanist common theory is advancing under cover of a seemingly radical ontological claim about community life. Displacing the distinction between productive labor (which produces surplus value) and non-productive labor (which circulates surplus value and facilitates the realization of profit), life itself is rewritten as simultaneously "productive" (thus fundamentally "creative") and "resistant" to capital's effort to exploit, capture and even measure it. This is the case because, according to the new communists, the commoning of culture coincides with the "death" of the labor theory of value. Negri writes, "intellectual, immaterial, and affective production (which characterizes post-Fordist labor) reveals a surplus. An abstract temporality—that is to say, the temporal *measure* of labor—is incapable of understanding the creative energy of labor itself."[37] Not only is the creative force of knowledge the new source of value, but it exceeds any capacity to calculate it. "Abstraction" is interpreted here not in materialist economic terms, as the reduction of "the different kinds of labour embedded in the different kinds of commodity to their common quality of being human labour in general,"[38] but in terms of the (cultural) *effects* of economic abstraction: as rationalization and regulation that seeks to control life by homogenizing it. Abstraction is treated more as an effect of a way of thinking, not as a material process.[39]

Value, however, is determined not by the rational machine of capitalist calculation, much less by "life force" of bodies, as new communists suggest. Calculation is itself an effect of the economics of value. Value is determined by the labor time necessary to produce a product (the law of value). The law of value does not end with the emergence of immaterial labor; it continues on "behind the backs" of the workers and determines their wages and the rate of profit. What is at stake in declaring the death of the labor theory of value is really a transvaluation of value *at the level of ideas*, or the desire expressed by Lazzarato "to change the principles of valuation, to change the way of conceptualising 'the value of value,' as Nietzsche suggested in *On the Genealogy of Morality* over a century ago."[40] By blurring the line between productive and unproductive labor and turning all labor (indeed all creative activity) into value-producing "life" subject to "regulation" and "control," new communist theory obscures the exploitative roots of ("Fordist" and "post-Fordist") capitalism in production. It also becomes the basis of rejecting as "abstract," "fixed," and "essentialist" precisely those theorizations that seek to address the abstract material processes that structure social life.

As for Virno and Lazzarato, so for Hardt and Negri, the automation of production and the transition of the global capitalist economy from the production of material goods to the immaterial commodities of "ideas, information, images, knowledges, code, languages, social relationships, affects, and the like"[41] is said to have dissolved property and, in turn, the material basis of class unity. In place of the working class whose existence is the objective consequence of having, in the words of Marx, freedom in a double sense, "that as a free individual he can dispose of his labour-power as his own commodity, and that, on the other hand, he has no other commodity for sale, i.e., he is rid of them, he is free of all the objects needed for the realization of his labour-power,"[42] Hardt and Negri, as I mentioned earlier, argue that the new figure of revolution is the multitude. Significantly, they insist that if "multitude is a class concept"[43] at all, it is *only* as a contingent and elective affinity. The multitude is a post-class concept of class, not as a relation to the means of production, but as a contingent "deployment" or a "project"[44] whose ever-changing conditions exceed the "older" forms of working class political movements aimed at collective ownership of the means of production. Class as a relation to property, this familiar argument goes, is too "narrow" and does not take into account that material workers have been largely superseded by immaterial workers, nor does it address the very different lifestyles and consuming habits among laborers that speak to emergent forms of common sharing. The concept of the multitude aims to substitute a multiplicity of "singularities" for what is represented as an "essentializing" and "fixed" notion of class "identity." Singularity, as Anna Curcio and Ceren Özselçuk propose in their introduction to a special issue of *Rethinking Marxism* on the common, is neither "abstract sameness" nor "atomic difference" and "breaks with the very terms of the question of the part and the whole, a question that presumes unity even before posing the problem of the relation."[45] The Marxist concept of working class as a *relation* is an implied target here. A dialectical concept that addresses both the vast differences in kinds of labor and products of labor in which the proletariat is involved as well as the underlying relationships that connect workers beyond these differences and that brings to light their common class interest in abolishing private property, "working class" in effect is represented as reducing all members of the working class to "abstract sameness." Indeed, this is often the "totalizing" ground upon which Marxism is rejected on the post-Marxist left: that is, for repeating what capitalism does to workers—reducing them to the "same"—when what needs to be affirmed, it is claimed, is their heterogeneity and autonomy.[46] However, these arguments themselves simplify the concept of working class and are deeply influenced by the ideology of the individual. At the same time, to establish left opposition to capitalism, Curcio and Özselçuk also distance themselves from the notion of "individuality" ("atomic difference"), which of course is not only discredited as humanist but as overtly reflecting bourgeois class interests. Singularity thus becomes a fluid "third way" identity in the common. It is the contingent space of the "in between," the "yes" *and* the "no."[47] Along similar lines, Hardt and Negri write that the common is "neither private nor public, neither capitalist nor socialist—and opens a new space for politics."[48]

But singularity merely "opposes" capitalism without negating capitalism. It opposes the effects of capitalism at the level of ideas—i.e., atomic difference—while rejecting the analysis of and struggle against capitalism for which putting an end to private property relations is in the essential (historical) class interests of working people. Singularity, to put it more sharply, is the sign under which the dominant left "holds out" for the possibility that it may not be necessary to abolish the inequality-producing system of wage labor to bring about the equality they claim to champion.

Within the post-property, post-class framework of the multitude, the common becomes the ground of the new subjectivities that both disrupt older binaries of identity (e.g., self/other, worker/owner, mind/body, human/animal, human/machine, capitalism/socialism) and create excessive forms of life resistant to the domination of the state and the appropriation of capital. To put this another way, the agent of production in the common is the posthumanist producer of common life which is always already social—it is a creative "force" that can't be reduced to the ("humanist") individual. From this standpoint, the individual's labor is "beyond measure" because it is "entangled" in so many other kinds of labor that resist calculation. At the same time, "immaterial" is equated with the social and the fluid (actually the "spiritual" as I discuss below), and "material" is made synonymous with the outdated, the fixed, and the isolated. The argument that global capital (still) directly exploits laborers at the point of production is, in effect, represented as a "humanist" nostalgia for the outmoded notion of the individual, material producer. The mathematics of measuring surplus value, consequently, disappears in the era of the excessive aesthetic cyborgian intelligence of the algorithm. The "widespread presence of algorithms in the daily life of digital culture," Tiziana Terranova proposes, are not only part of a new digital automation that "unfolds in networks consisting of electronic and nervous connections so that users themselves are cast as quasiautomatic relays of a ceaseless information flow."[49] They also have a "capacity to express not just 'use value' as Marx put it, but also aesthetic, existential, social and ethical values."[50]

In actual fact, the development of knowledge workers and diverse forms of consumption in (industrial and postindustrial) capitalism have never signaled the end of capitalist private property relations but rather the increasingly global expansion of wage labor, within which the gap between exploited and exploiters ever widens. But such changes are the touchstones of cognitive capitalism, and they are used to declare new developments *within* capitalism as transformations of the very "identity" of capitalism itself. On these terms, the concept of the multitude "opens" class to a much wider range of new practices (such as open software and community gardening) that, it is argued, radically challenge capitalist relations and do away with the idea of a working class political party aimed at taking control of production as necessary for the socialist transformation of capitalism. For common theory, communism is here and now.

One of the reasons commonist readings of capitalism have become so influential is that they turn the contemporary ethical consumer branding of products into a radical postcapitalist politics. The ethical subject sees herself as participating in ways of being with others that combat the privatizing forces of capitalism—whether through community gardens or credit unions or taking part in digital communities or "exodus" from the "grid" itself—even though all aspects of her life (including consumption) remain determined by the relations of wage labor. While these practices may resist certain forms of alienation, they in no way challenge capitalist relations of wage labor. Such practices instead foster "ethical capitalism" and are often easily accommodated and appropriated by capital.

Uber, to take one example, was initially heralded as a radical new form of the "sharing economy," since it creates a driving service outside of both private taxi services (monopolized by a few companies, and able to charge monopoly fares) and public transportation (run by the state and increasingly expensive). Its workers are "independent contractors," essentially freelance laborers who arrange to pick up passengers through its online network. Uber drivers use their own cars, and create their own schedules and are rated by their passengers. But what is really at stake is unfettering the company from any obligations to workers—whether in the form of a regular wage, or worker's insurance and protection on the job. Workers not only have to have a car in the first place, but are not compensated for car repair costs, are liable for accidents, and are not guaranteed work. The sharing economy is actually the corporate fantasy of the worker to whom the company is obligated to provide nothing, and whose customer reviews in many ways determine which workers keep their job, which workers will be able to pay their rent. This of course is translated into a new form of consuming culture that "breaks" from traditional cultures of work. In its consumers, it fosters the notion that what they are participating in is a new kind of (digital) community in the city, that what they are paying for is outside the normal circuits of exchange. In actuality, it legitimates austerity and anti-worker policies and perpetuates the market ideology that social priorities should be set by (private) desire not collective need.

Jodi Dean has critiqued the way in which the dominant left in the last several decades has uncritically celebrated consumer culture and focused on "micropractices" that "depoliticized into lifestyle choices" economic alternatives to capitalism and ultimately "fetishize" "the local."[51] Challenging the overwhelming left opposition to the very notion of a working class party, Dean makes an important argument for the need for building a working class party that is capable of helping people see the underlying class relations that connect them through their differences in jobs, lifestyles and consumption. However, in a sign of how seductive the (class) discourse of the knowledge economy is among left thinkers, even Dean remains trapped within the very cognitive capitalist framework that informs the scholars and activists she critiques. For instance, Dean writes that the problem with Negri is that he didn't fully identify contemporary capitalism's "new forms of capture and subjection."[52] In other words, for Dean as well as for Negri and left theory more generally, capitalism is above all a regime of "control": "Our mediated sociality affirms the structures of capitalist control more deeply and completely than the institutions of civil society ever did—especially under circumstances that amplify injunctions to individuality."[53]

In positing the end of capitalist exploitation of labor through automation, theories of the "communism of capital"—which read the central conflict in capitalism today as the capturing of the common produced by affect and immaterial labor—erase the material relations of social transformation. In effect, they put forward an idealist theory of the common that affirms the radical force of spontaneous, voluntarist feeling as the basis of change. And perhaps one of the most telling extensions of such arguments is advanced by the advocates of "disaster communism" who propose that "disaster communities" (e.g., the cooperative, non-state, non-market practices people spontaneously engage in when confronted with disasters) "give us ample reason to believe that local, emergent bricolage can efficiently meet human needs even under the most adverse conditions."[54] This is a sentiment shared by Naomi Klein, who writes in *This Changes Everything*, "there is no reason why future disasters cannot be laboratories for those who believe in reviving and reinventing the commons, and in ways that actively reduce the chances that we will all be battered by many more such devastating blows in the future."[55] Regardless of intentions, these arguments not only make a (posthumanist) virtue out of disaster, but they obscure the relation of such communities to the austerity measures of capital—which are aimed at lowering the cost of production (of the laborer), forcing the laborer to cover with their wages more aspects of reproduction that were once funded by a part of the collective surplus value. Advocating such communities actually lets capital and the state off the hook, since the more "people themselves" are encouraged to help each other without the help of the state (by giving to charities, engaging in ethical sharing, etc.) and "outside" of the market, the less people come to demand higher wages or better social services, and the less likely they are to challenge the system that systematically leads to disaster conditions of life. Such arguments also feed the reactionary narratives that promote reduction of consumption among the working class as the answer to ecological crisis (itself a product, not of humanity, human reason, or human consumption, but historical, and therefore changeable, class relations, above all). This partly explains the appeal of such arguments to religious figures, like Pope Francis, who invited Klein to the Vatican to support his call for a more egalitarian, ecological capitalism. The issue here is that what is represented as "radical" and as supportive of "communism" is a way of updating capitalism ("posthumanizing" it) so as to forestall more radical change.

The Material Roots of the Common

Although, following autonomist Marxists like Negri, the departure for virtually all new communist theory is the reading of Marx's *Grundrisse* as a tutor text in cognitive capitalism, in fact, in both the *Grundrisse* and *Capital*, Marx theorizes cooperation as a material force of production. Labor has always been a productive force, he explains, but its cooperative force is unleashed in capital's social relations of production. "A large number of labourers working together, at the same time, in one place [. . .] in order to produce the same sort of commodity under the command of the same capitalist, constitutes the starting point of capitalist production. This is true both historically and conceptually."[56] This cooperative force produced by workers is not, as new communists like Negri claim, an "autonomous" force that exists outside of capital, but a direct result of workers' status as propertyless producers who, as such, are forced to work for capital to survive. "[T]he co-operation of the wage-laborers," Marx emphasizes, "is entirely brought about by the capital that employs them."[57] This is because the "objective of all development of the productivity of labour within the context of capitalist production is the shortening of that part of the working day, in which the worker must work for himself, and the lengthening, thereby, of the other part of the day, during which he is free to work for nothing for the capitalist."[58] Capital and labor exist in a contradictory relationship and the wealth of capital grows in proportion to the private appropriation of surplus value produced by the collective worker. As a result, no matter how global the socialization of production becomes, and no matter how much capital comes to rely on social knowledge in the production process, "the positing of social labour in the form of the contradiction of capital and wage labour—is the ultimate development of the *value-relation* and of production resting on value. Its presupposition is—and remains—the mass of direct labour time, the quantity of labour employed, as the determinate factor in the production of wealth."[59] This gives rise to the great struggles over the working day, as workers fight to limit—and eventually end—the theft of value by the capitalist and to transform the relations of private property.

Contemporary theories of the common take the effects of the socialization of production under capitalism as a sign of a fundamental change in capitalism that severs the production of "common" products like language, knowledge, affect, and so on, from their material relation to exploitation in production. In doing so, commonist theories not only isolate knowledge from direct production, thereby erasing the implementation of knowledge always involved in manual labor. Just as significantly, in their effort to treat only "immaterial" labor as "creative" and to disconnect it from the exploitative relations of capitalism, digital products and services are represented as somehow autonomous from the very structures on which they depend. On this basis, "resistance" to capitalism is increasingly seen as a form of "exodus" from capital, and not as the class-conscious fight to transform it. But Marx's point is that the expansion of capitalism that brings workers from all over the world together in the production process *also* shatters that cooperation by forcing workers to compete against each other and increasing their productivity through technology. Through introducing technology into the production process, capital continually assembles workers for the purpose of appropriating higher levels of surplus value. Bringing people together in common to work in and of itself, as a result, is not at all opposed to capital, as commonists suggest. And nor is common "consumption" outside of work. These are really forms of the "communism of capital" that the new communists claim to oppose. They represent as "radical" technological and cultural changes within existing property relations. Revolutionary communism is instead committed to "the most radical rupture with traditional property relations."[60]

To put this another way, if early apologists of capital represented capital as the spontaneous producer of surplus value (in the circuits of exchange), common theory produces a similar mystical effect in its representation of workers as producing value cut off from direct production (in the circuits of consumption and culture). Immaterial laborers, it is argued, are the new class that possesses a mystical "potential" outside of any property relations; again, it is now "pure subjectivity"[61] that creates value. Such claims erase the roots of surplus value in the exploitation of direct labor, and lead therefore to the same ideological place as the defenders of the miraculous capitalists: to the claim that workers do not need to transform capitalism. The difference is that, whereas defenders of miraculous capital suggest capitalism is "the end of history," the dominant left today suggests that capital's immanent movements lead to its own transformation into a more equal society with the advent of the immaterial labor of the common. In the words of Robin Mackay and

Armen Avanessian, "the only thing that will kill capitalism is its own 'excess.'"[62]

Mackay and Avanessian draw their undialectical conclusion in their reading of Marx's analysis of the organic composition of capital: the ratio of constant capital (capital invested in technology and the means of production) to variable capital (wages of labor). For Marx, as the role of technology expands in the process of production, capital has a tendency to invest less in labor relative to technology that increases the productivity of labor. And, since labor power is the basis of surplus value (profit), the less capital invests in variable capital (workers) and more in technology, the more capital's tendency toward accumulation creates conditions for crisis: specifically, the falling rate of profit. Mackay and Avanessian, following Negri and the school of post-*operaismo*, translate this into a fantasy in which capitalism, no longer rooted in the direct exploitation of labor, will succumb to its own undoing without the need of any action of workers taken against exploitation. They write, "Since the 'new foundation' created by integrated machine industry is dependent not upon direct labour but upon the application of technique and knowledge, according to Marx it usurps capitalism's primary foundation of production upon the extortion of surplus labour. Indeed, through it capital 'works toward its own dissolution.'"[63] On these terms, capital now depends on "integrated machine industry" that itself depends on the increasing expansion of (social) knowledge and technology (not the direct extraction of surplus value in production); investment not in living labor but in the technology (dead labor) that leads to "common" knowledge undoes capitalism from within.

However, this is a telling misreading of Marx, and I address Mackay and Avanessian's argument because it represents the dominant reading of Marx by accelerationists like Mackay and Avanessian and the new communists alike (despite immanent differences between them[64]). The "rationalization" of production by scientific means, for Marx, does not itself abolish the exploitation of labor. He actually argues the reverse, that exploitation "intensifies" the more productive labor becomes[65]—and that those who fail to grasp this are prone to the "mystification implicit in the relations of capital as a whole" that attributes the productivity of labor to the "*productive power of capital*."[66] While the falling rate of profit that stems from greater investment in technology over labor leads to ever more global economic and social crises, there are countervailing tendencies that can counteract this (like massive destruction of existing stock, to restart the recovery process). But, more importantly, it is only through the self-conscious organization of the working class against capital that capitalism can actually be transformed into a new mode of production.

When Marx argues that, "As co-operators, as members of a working organism, [the working class] merely form a particular mode of existence of capital. Hence, the productive power developed by the worker socially is the productive power of capital,"[67] he in fact demonstrates the illusory nature of the idea that capitalism will die of its "own excess." The social knowledge created within capitalist production is "a particular mode of existence of capital" fundamentally rooted in the dialectic of labor and capital. As such, it is not the social knowledge (the "common") that will lead to the end of capital and the development of radically new social relations. Sharing products that have been produced through exploited labor does not undermine exploitative production. Capital in fact depends on workers' sharing of products of consumption. Not only do the policies of austerity make such sharing outside of the market and the state increasingly necessary for workers to survive (which is why charities, churches, food pantries, and the sharing economy have become so much a part of American life since the 2008 recession), but the role of the family unit in capitalism has always been a key site for reproducing the working class, allowing several people to survive together on low wages by sharing objects of consumption.[68] Valorizing common consumption and culture as radical serves to hide the foundation of capitalist society and forestall the collective struggle by the working class for a truly free society: a society in which no one can exploit another.

To summarize, "cognitive capitalism," far from representing capital's "self-destruction," represents an increase of the rate of exploitation. The immaterial commons of social media have consolidated rather than minimized or abolished class inequality and this is because labor is still socially required to make itself exploitable by capital: workers must acquire the technical and consciousness skills necessary for high-tech capital in order to survive and must do so at a level of technical development that is less labor intensive and so requires less workers. The social media commons is where the flexible labor force of cybercapitalism learns to cooperate in the borderless factory by imagining itself as above and beyond the class binaries of industrial capitalism and its crude class solidarities. In reality, and as is increasingly difficult to avoid, intellectual labor is steadily subject to the same regimentation, calculation and exploitation that cognitive capitalist theory claims exists only in material production. As one article explains,

> A new generation of workplace technology is allowing white-collar jobs to be tracked, tweaked and managed in ways that were difficult even a few years ago. Employers of all types—old-line manufacturers, nonprofits, universities, digital start-ups and retailers—are using an increasingly wide range of tools to monitor workers' efforts, help them focus, cheer them on and just make sure they show up on time.[69]

Monitoring worker's efforts, of course, is a euphemism not for control (which is how left theory translates such phenomena), but for making workers more productive—getting more from the value of the labor the owner paid for—and, therefore, increasing the rate of exploitation. This is ultimately what is at stake in the revelation about Amazon, which, as a *New York Times* exposé on the working conditions there made clear, "is conducting a little-known experiment in how far it can push white-collar workers, redrawing the boundaries of what is acceptable."[70] Which is another way of saying that immaterial labor is increasingly proletarianized along the lines of material labor. The report indicates that "Amazon is in the vanguard of where technology wants to take the modern office: more nimble and more productive, but harsher and less forgiving"; "fluid but tough, with employees staying only a short time and employers demanding the maximum."[71]

Instead of representing the "end" of capital's former class relations of exploitation, fluid, contingent, and flexible conditions of labor (e.g., Negri's contingent class "projects") are fundamental to contemporary capitalism's regime of accumulation. The stark lines of class relentlessly reassert themselves. The ideological role of left theory is to bury them in the haze of heterogeneity.

One of the central issues here is this: theorists for whom capitalism is cognitive capitalism constantly reference the "disappearance" of manual labor and manufacture ("material labor") in contemporary capitalism, and they endlessly champion the role of knowledge in creating the world. But these same theorists conveniently ignore the role their own theories play in both making today's material labor *invisible* and in fostering "differences" between immaterial and material wage laborers—the very differences produced in the labor process under capitalism and that get reified by bourgeois ideology into class-as-lifestyle politics or class as "symbolic capital" or class as classy, in order to foment competition among workers, reduce wages, and prevent any serious resistance to capital. The lack of resistance to capital (the deterioration of the labor movement) is then referenced as "confirmation" that the working class is dead and that a new class has emerged that needs to be "liberated" from the "fixed" binaries of the past. Left theorists, in other words, ideologically sever the connections in the social and rewrite them as a new posthumanist ontology of contingent common being. So eager is the left to abandon the working class and its earlier political struggles to end exploitation that some, like Mikkel Bolt Rasmussen in *Crisis to Insurrection*, come close to celebrating the "ruined" state of the working class in spite of the devastation this has brought about worldwide: "it might not, however, be a drawback that the political forms of the Western working class are in ruins. We need to move elsewhere anyway. The twentieth century is a testament to that."[72]

The "Supernatural Nonsense" of the (Digital) Petty Bourgeois

New communist theories, it is important to note, ignore the fact that "common" knowledge remains bound up in exploitation under capitalism, which divides workers between low and high tech sectors in the division of labor. In effect, the new communist privileging of knowledge ends up uncritically reinscribing these divisions, by first separating "knowledge" from labor—as if knowledge is not always involved in productive labor—and then privileging the sector of knowledge workers as the basis of all value today. Uncritically accepting the commodified forms of capitalist life, they regard workers as a separate class on the basis of *what* they produce, not on the basis of the relations of production and their *relation to the means of production*. Rather than see intellectual workers as part of the same class as material workers, theories of cognitive capitalism oppose both big capital as well as the working class. This self-distinguishing from material labor and the intense focus on singularities in the common are often represented as a sign of the common's commitment to difference and heterogeneity. In actuality, such concerns are key elements of the petty bourgeois framework of the new communist theory.

The petty bourgeois historically reflect the class interests of the small producers—whether small owners of land (peasants) or small (often family) businesses—those whose conditions of life and property capital is continually destroying, as capital's tendency toward concentration and monopoly increase. The writings of the new communists and the theory of cognitive capitalism more generally reflect a new articulation of the petty bourgeois in the digital era of capitalism. As Teresa Ebert and Mas'ud Zavarzadeh argue in their critique of new communist theory, cognitive capitalism "in many ways is a re-appearance of the 'putting-out' system of labor, the regime of small producers."[73] New communist writings are, in other words, "the expression of the class interests of the new petty bourgeoisie—the intellectual workers—whose class interests are antagonistic not only to the proletariat but also to large-scale capitalism."[74]

Such intellectual workers are often highly skilled and educated, own their own technology, and contract their work out to larger companies. They may, for instance, work on video games, or produce software like apps, or come together with others for a particular, short-term project. On the one hand, therefore, they have more freedom in their working conditions and a great deal more autonomy relative to wage laborers. Their work is "contingent" and especially "project"-oriented—precisely those qualities that are so privileged in new communists' representation of the common. On the other hand, they lack security and are under constant threat of having their work further automated and taken over by either other petty bourgeois or wage-laborers who can produce the same products cheaper. As small property owning producers, they oppose capital insofar as it threatens to ruin their small property or limit their own individual freedom to work autonomously and creatively on their own terms, but do not identify with the (propertyless) proletariat and reject the struggle to abolish capital, since they benefit from the existing property relations, as precarious as their own position may be. (In this sense too, the new academic concern with the "precariat" is more a concern with the class status of the petty bourgeois knowledge workers, who are becoming increasingly proletarianized in austerity capitalism, than it is about the situation of the working class, the majority of whom have always lived precarious lives.)

The new communists, in other words, are the theoretical wing of the digital petty bourgeois; they universalize their particular freedoms and translate these class interests into the theory of the common as the space of both of working together free from the instrumentalities of the market and the state as well as the space of singularities. For this reason, Ebert and Zavarzadeh argue,

> The opposition to capitalism in the revolutionary vocabularies that Negri, Žižek and other 'new' communists use to describe their antagonism to capitalism is only the opposition of the new petty bourgeoisie to big capitalism. It is not an opposition to capitalism. 'New' communism and its cognitive economics are reformist projects to secure a place for intellectual workers within existing capitalism and not to overthrow the system.[75]

Negri, in fact, has become quite explicit about the class interests he represents: "With all due respect to those who still comically believe that revolutionary possibilities must be linked to the revival of the working class of the twentieth century," he writes, today "we are still dealing with a class, but a different one, and one endowed with a higher power. It is the class of cognitive labor. This is the class to liberate, this is the class that has to free itself."[76] Negri's argument is of course revealing. He characterizes the class of immaterial laborers as fundamentally different from material workers—even invested with a "higher power" relative to them—because, he assumes, they alone are employed in jobs which require the use of knowledge and skill. This repeats the familiar representation of immaterial labor in cognitive capitalism, and suggests that material workers, who work in direct production and many of whom work in the global South, are the class of the past. His interest in class "liberation" extends to a privileged sector of intellectual workers who benefit from capitalist property relations but desire a more ethical distribution of wealth. As another analysis of Negri and Hardt's writing puts it, the left subjects for whom new communists like Negri write "want a theory which explains their struggles, which are not struggles for bread and butter. Negri and Hardt fit the bill."[77]

The position of petty bourgeois knowledge workers between big capital and the proletariat, not surprisingly, leads them politically and philosophically to a whole range of "in betweenisms" that resist committing to either side in the class struggle between capital and labor, while remaining "open" to alliances with both. It is this which Lenin identified with the class "opportunism" that the working class struggle continually fought against in the period leading up to the 1917 Revolution. But of course new communist theorists situate themselves as exceeding such categories as working class, petty bourgeois, and capitalist. As I have argued, they are quick to reject as "essentialist" the classical concepts of Marxism and the economic concepts necessary to understand exploitation. Yet, one of the important implications of such theories is that their "anti-essentialist" approach to class leads to a spiritualist theory of class and the common that actually reinstitutes the very binary logic it aims to avoid. This is the double-move of new communism.

In the first move, for instance, theorists of the common like Gigi Roggero argue that "there is no idea of an original unity of labor that is then divided and alienated by capital and therefore in need of recomposition, nor is there a concept of consciousness that must be revealed to rejoin the class-in-itself with the class-for-itself."[78] Instead, the only space of biopolitical resistance to the "communism of capital" exists from within the rupturing of subjectivity itself, making transformation of the relations of wage labor unnecessary. Where do such anti-essentialist theories of labor and the common lead? Following the cultural studies dogma that class is "performative" rather than a relationship to the means of production, and thus that class cannot be reduced to an essentialist "identity," new communists treat the common as spontaneously produced in the struggle for the common. "What is common, what can be described as unity," Leon de Mattis insists in "What is Communisation?," "stems from the struggle itself, not from our identity within capitalism." And this is the case, "because no particular condition can any more identify itself with a general condition."[79] Precisely for this reason, the problem of earlier revolutionary politics is that "they comprehended communist revolution as the victory of a class over another class, not as the simultaneous disappearance of classes."[80] This led them to "the traditional idea of a transition period during which the proletariat, once victorious, assumes the management of society for an intermediate period."[81] This argument (which is ultimately an argument against socialism) is developed further by others, like the writing group and journal *Endnotes*, to dematerialize labor itself. "On this [classical Marxist] view labour is something that exists quasi-naturalistically in the product, and exploitation is seen as an issue of the distribution of that product—thus the 'solution' to capitalism is seen as workers, via the state or other means, shifting that distribution in their favour."[82] New communist theory, therefore, "calls into question any political conception based on the affirmation of the proletariat as producer of value,"[83] since this simply reinstates the very kind of "essentialist" approach to class that new communists seek to put under erasure. Having entirely erased the material conditions of class, labor, and the common, and rejected as "quasi-naturalistic" the economic theorization of value as determined by labor time, "communism" becomes a utopian "event" in which all classes and contradictions are instantaneously destroyed: "we are now in a position to understand that communism can only be the simultaneous disappearance of social classes, not a triumph, even transitional, of one over another."[84] Communisation is the "moment" of

> the immediate production of communism. By communism we mean a collective organisation that has got rid of all the mediations which, at present, serve society by linking individuals among them: money, the state, value, classes, etc. [. . .] Communism will thus be the moment when individuals will link together directly, without their inter-individual relations being superimposed by categories to which everyone owes obedience.[85]

For the new communists, even such spiritual readings of communism as Jacques Camatte's claim in *Capital and Community* that "communism is neither a mode of production, nor a society"[86] is too material. The new communists want nothing less than a communist rapture.

The new communists' resort to spiritualism is a symptom of what Lenin calls in *"Left-Wing" Communism: An Infantile Disorder* the "supernatural nonsense" that petty bourgeois communism continually produces in its substitution of left-wing "childishness" for scientific socialism.[87] Characterized by, for instance, a stubborn refusal "to understand [. . .] the necessity of a strictly objective estimate of all the class forces and their inter-relation in every political action"[88] and a fetishizing of "individual" acts of resistance,[89] the "petty-bourgeois revolutionism" was ultimately one of the enemies of the working class movement in Russia prior to the revolution.[90] Echoing the liberal populist rejection of working class leadership, it opposed the role of the Party and Party discipline in the struggle to expropriate the expropriators, which "is equivalent to disarming the proletariat in the interests of the bourgeoisie. It is akin to that petty-bourgeois [diffuseness], instability, incapacity for steady, unified and harmonious action, which, if given encouragement, must bring to nought every proletarian revolutionary movement."[91] Left commonism is the "infantile disorder" of the petty bourgeois in fear of their transfer into the proletariat by the dominance of monopoly finance capital. The petty bourgeois—who imagine their particular class interests as cultural workers as the source of true freedom—see in their dissolution by big capital and transfer into the proletariat the end of civilization as such. It is not a coincidence that the more capitalism seems to teeter on the brink of collapse, the more "disaster" becomes fetishized in left theory. What becomes in posthumanist theory the ontological condition of "life itself" is an effect of the ideological universalization of the class interests of the petty bourgeois and essentially their desire to be "delivered" from the material contradictions between wage labor and capital upon which their existence as a dependent class rests.

In fact, the knowledge-created common of cognitive capitalism becomes itself equated with the (inherently) frictionless space that exceeds all boundaries: what Curcio and Özselçuk call a "substanceless substance."[92] The central premise of the common is that "what is shared is not a substance that is natural, fixable, or given, but one that is to be produced and re-produced and one that is open and multifarious. In fact, one can conceive of this as a 'substanceless substance.'"[93] The "immaterial" here is once again associated with the boundarylessness that comes from the fundamentally "common" nature of knowledge products created by the immaterial worker and his/her affect, languages, etc. In fact, so "substanceless" are knowledge and immaterial labor understood today by those theorizing the common that they take on spiritual characteristics that do away with all (material) conflicts and concepts:

> Spirit, like intellectual or immaterial labour, has a tendency to cross the borders; it is without spatial existence and does not reduce to its manifestations. In the era of immaterial labour and cooperation between minds it is not possible to think social conflicts in terms of the friend/enemy dichotomy or in terms of the conflict between two classes, nor in terms of liberal (private/public) or socialist (individual/collective) traditions.[94]

But—and this is the second move in the double-move of new communism—while the common of immaterial labor is represented as a break from all binary thinking, it is itself actually inscribed in (and therefore reasserts) a whole series of binary relations. After all, the "immaterial" is equated with boundaryless fluidity, the "material" is associated with the "bounded" products of industrial capitalism. While the commonists claim (as always) to be "beyond" binaries, they institute a new set of oppositions that reify the division between material and immaterial workers, the latter of whom are represented as the new revolutionary class. Industrial labor becomes the ground of "fixed capital" and "dead" labor, while postindustrial labor is embraced as variable and relative (constantly changing) space of "living" labor and vitality. The material worker whose labor is exploited in direct production is represented as a relic, whose struggles against capital represent a now "outdated" struggle, while the creative class is the "new" class of "liberation." To say it again, new communism (and its new binaries) reflects the "higher" class interests of petty bourgeois knowledge workers today who, in the face of rapid proletarianization brought about by the falling rate of profit, identify not with the "revolutionary class [of the proletarians], the class that hold the future in its hands," but with the bourgeois interests that seek to eternalize the (capitalist) present.[95]

The Paris Commune and the Collectivity of Labor

What is ultimately at stake in the petty bourgeois rejection of class for the common among new communists, I argue, is the question of the transition to communism. Regardless of their radical rhetoric and self-distancing from the "communism of capital," all ultimately embrace the "spontaneous communism" that already exists within capitalism, or they offer meditations on the event of communism to come (e.g., Douzinas and Žižek's *The Idea of Communism*) that bypasses socialism in favor of a miraculous communism without revolution.

It is for this reason that the Paris Commune of 1871 has become the focus of new communist and commonist writing more generally. The Commune represents an unprecedented effort by an organized working class to take over the state to collectively produce a radically new set of social relations in the interests of the working class. What is particularly telling is the way that the common becomes used in new communist writings to turn the Commune into a cultural space of ethical living or a mystical event that justifies the rejection of the working class party. That is to say, dominant readings of the Commune all adamantly reject the dictatorship of the proletariat without which, even in its still undeveloped form, the Paris Commune would not have existed much less survived two months, before being brutally exterminated by the full force of the state of the French and Prussian ruling classes in a virtually unprecedented alliance between two states that had been at war less than a month before.

Prior to March 1871, the courageous and self-sacrificing working people of Paris had been devastated by nearly six months of fighting the Prussian army, to which the French bourgeoisie would capitulate (signing a peace treaty in February 1871) against the widespread opposition by Parisians, who were prepared to fight to the end to resist Bismarck's invasion. When the French military came on March 18 to disarm the working class in Paris, because French leaders in Versailles feared that the people's refusal to accept the armistice with Prussia would lead to a revolution against the ruling class, the workers—led by women—refused to give up the cannons and declared the Paris Commune, which included soldiers in the National Guard, small factory workers and craftsmen, trades workers, petty bourgeois owners, intellectuals, and journalists. Within days, workers began to institute reforms made possible by taking control of, and beginning to transform, key elements of the state. First and foremost the Commune revoked the standing army and established the National Guard as the people's army, and instituted open elections for all government posts, stipulating that those filling government posts would earn workers' salaries and be subject to recall.

Having thus begun the transformation of the ruling class state into a proletarian state, the Commune went on to institute a wide range of radical practices, demonstrating the great "historical initiative" of the masses.[96] It made the police answerable to the Commune; remitted all payments of rent; planned for the control of factories by workers; set up free schools from which religion was excluded; developed a women's union which demanded social equality for women; abolished night work; and intervened in national chauvinism by electing leaders of the Commune from numerous European countries. Its many achievements, Marx writes, "show that it was a thoroughly expansive political form [. . .] the product of the struggle of the producing against the appropriating class"[97] that smashes all "ready-made utopias" by "set[ting] free the elements of the new society with which old collapsing bourgeois society itself is pregnant."[98]

Marx, Engels, and Lenin therefore celebrate the Paris Commune as the first working class struggle to take control of and begin to transform the state, and to institute by the people themselves policies and programs directly in the interests of the proletariat. It was, Marx says, "the dawn of the great social revolution that will liberate mankind from the regime of classes forever."[99] At the same time, Marx and Engels insist on the importance of learning from the mistakes made by the leaders of this first working class movement to take control of the bourgeois governing institutions. After their victory on March 18 against the bourgeois military, for instance, the National Guard—the workers' army—was in a position to quickly launch a counter-attack against Versailles, where leaders of the French military and heads of state, like Adolfe Thiers, had fled. Yet rather than attempt to overthrow the French state, the Commune remained in Paris, where it was isolated, and eventually surrounded by an army that, in effect, had been given many weeks' time to prepare for the looming civil war against the Parisian workers. The leaders of the Commune, in not "at once marching upon Versailles, then completely helpless," allowed Thiers's army to return with "revolting" violence.[100] On May 21, 1871, two months after taking control of Paris, nearly "200,000 hostile troops [bore] down on one tenth of that number of Federals" fighting for the Commune.[101] The Parisian working class were defeated within days. This however did not stop the army of the French State from conducting in subsequent weeks such an indiscriminate massacre of working class men, women, and children throughout Paris—estimates range from the "gross underestimate of 6,500" to the more realistic figures of "20,000 to 37,000"[102]—that even Versailles apologists were forced to publicly distance themselves from the state-sanctioned class terror and slaughter. Nor did the leaders of the Commune move to take control of the bank, which would have drastically limited the ability of the Versailles government to function, especially curtailing its military resources and the scope of its attack on Paris. This was, Engels points out, "a serious political mistake," moreover, since had the bank been in the hands of the Commune, "It would have meant the pressure of the whole of the French bourgeoisie on the Versailles government in favour of peace with the Commune."[103] In failing to take control of the state's finances, the Commune left one of its most critical levers in the hands of the ruling class. "But what is still more wonderful," Engels continues, "is the correctness of so much that was actually done by the Commune."[104] It was the momentous achievements of this first of its kind commune, on the one hand, and, on the other hand, the limits that sprung from the struggle's lack of historical precedent that led Marx to insist on the critical significance of the dictatorship of the proletariat in the struggle for communism. "Do you want to know what this dictatorship [of the proletariat] looks like? Look at the Paris Commune."[105]

Whereas Marx begins his analysis of the Commune on the dialectical and materialist basis that "the soil out of which it grows is modern society itself,"[106] and explains that the basis of class contradictions in the private ownership of the means of production signifies that "it was only the working class that could formulate by the word 'Commune' and initiate by the fighting Commune of Paris" the emancipation of the exploited population,[107] what makes the Commune significant for new communist writers is what Hardt calls "the common in communism."[108] Hardt and Negri argue, for instance, that although the Paris Commune may have been representative of the traditional modern conception of insurrection, "characterized by a movement from the insurrectional activity of the masses to the creation of political vanguards, from civil war to the building of a revolutionary government," today "such sequences of revolutionary activity are unimaginable" because "insurrectional activity is no longer divided into such stages but develops simultaneously" in "the flesh of the multitude."[109] On these commonist terms, what erupts in the multitude's spontaneous and everyday activity are spaces within capitalism for momentary ruptures in the existing cultural logic. Commonist readings of the Paris Commune, like Hardt and Negri's reading, renarrate the mantra that contemporary capitalism is radically different from industrial capitalism, and substitute for working class revolutionary collectivity a post-class, multitudinous theory of the common as the ground of the new posthumanist subjectivities that disrupt class binaries of worker/owner and that exceed the domination of the state and the appropriation of capital. It is claimed that the era of class revolution is dead, and with it socialism—without which communism is a utopian dream.[110] Which is another way of saying that new communists help to ensure there is no "building of a revolutionary government" by declaring its impossibility and undesirability. According to Negri (*Goodbye Mr. Socialism*) and Žižek ("How to Begin from the Beginning"), socialism is an invention of capitalism to perpetuate itself. Socialism, in other words, is a defense against capitalist crisis. It uses the central planning of the state to organize production. Such arguments essentially represent socialism as the social welfare state (an argument which mimics the right wing attacks on investment in social programs), in order to justify the left utopian fantasy of a communism that does not involve working people taking control of the state. If the Paris Commune has anything to teach us, such writers suggest, it is the significance of the production of the common, as a continual state of becoming that exceeds (class) definition but is continually produced anew without the need of revolution. But this is actually a "lesson" in conceptually and politically disarming the proletariat.

If there is any stronger indictment of this anti-socialist Marxism today, it is perhaps in the figure of Yanis Varoufakis, the former finance minister of Greece who in January 2015 came to power in Syriza's Left campaign to champion people's interests against the barbarity of austerity measures imposed on the Greeks by the IMF, the European Union, and the European Central Bank—only to end up accepting and negotiating the terms of austerity. A self-proclaimed "erratic Marxist," Varoufakis was a student at Essex while Ernesto Laclau, of one of the most influential theorists of post-Marxism, was a professor there, and it is not surprising that, as Dan Hancox contends, "Syriza built its political coalition in exactly the way Laclau prescribed in his key 2005 book *On Populist Reason*."[111] One of Laclau's main "lessons" is that class is an excessive, overdetermined set of processes irreducible to the "essentialist" binary of capital/labor, and along with Chantal Mouffe declared the "impossibility of society."[112] Like many new communists, Varoufakis claims (in his op-ed, "How I Became an Erratic Marxist") that while he agrees with many aspects of Marx's theorization of capitalism, he (conveniently) opposes the idea—fundamental to Marx's analysis—that the basis of capitalists' power is "reducible to their capacity to extract labour from employees."[113] He argues instead that profit also comes from capital's ability to exploit "a given community of consumers," repeating the commonsense of cognitive capitalism that value is created in consumption. But most significantly, in trying to explain why he did not and does not argue for a more "radical" set of proposals in Greece, Varoufakis claims that it was actually the "radical" politics of the working class in the era of Thatcher that led to the defeat of the left (an argument which entirely absolves the New Left from its own abandonment of working class politics in favor of social movements that disconnected social inequality from class) and that he is not going to make "the same error twice" by defending working class movements for "socialism." To support the working class, Varoufakis suggests, one has to protect the class that exploits it. This is rather like the "communists" who defend bourgeois society "for the benefit of the working class."[114] His resignation from Syriza—when, even in the face of a people's referendum overwhelmingly opposing more austerity, Syriza leaders announced plans to pursue an even more brutal austerity plan—should not deflect attention from the relation of his theorization of capitalism, class, and socialism and the powerful place he occupies in the institutions of European capitalism. The question is not why an "errant

Marxist" is not "more radical" having achieved a position of immense political power. The question is why capitalism is able to advance some of its most anti-worker policies in discourses that are virtually indistinguishable from the "anti-capitalist" and "new communist" left.

In spite of immanent differences with other new communists, Alain Badiou in *The Communist Hypothesis* also begins from the assumption that the Commune cannot be reduced to a class struggle. Badiou conceives the social as a conflict between the included and the excluded, those who are "counted," and those who are not. For him, the social is rigidly and arbitrarily structured by dominant forces of the state, which set the terms for those who are recognized as "existing" in that social world and those who are not (the inexistent) and consequently brutally subject to those forces. Badiou sees great struggles like the French Revolution, the Russian Revolution, and the Paris Commune as "events" or sites of "singularity" to the extent that they create "the existence of an inexistent."[115] Thus he writes that "18 March 1871" is "a singularity because it evokes its being in the appearing of its own multiple composition."[116] When the working class refused to give up their arms and protected their cannons from confiscation by the bourgeois military, this is not for Badiou significant as a class act, in which people defend what they have produced and fight for as a means of taking their lives into their own collective hands. Instead, it speaks (in mystical, phenomenological terms) to an "appearing" and the creation of a new "being," that is "the exigency of a new political appearing."[117] In this way, Badiou can talk about the violent conflict between the working people of Paris and the ruling forces as if this conflict exceeded class struggle and manifested instead a new, amorphous, and radically chaotic ontology.

According to Badiou, the problem with Marx's analysis is its emphasis on "social determination (workers)" that leads Marx to address the importance of the state in the transition to communism. "Marx proposed an account of the Commune," writes Badiou,

> that is wholly inscribed in the question of the state. For him, it comprises the first historical case in which the proletariat assumes its transitory function of the direction, or administration, of the entire society. From the Commune's initiatives and impasses he is led to the conclusion that the state machine must not be 'taken' or 'occupied', but broken.[118]

But for Badiou the real significance of the Commune is that it "broke" not only from the state and the Left but all representational politics, which are for him the embodiment of the state imposition of determinism. The Commune, he argues, is significant as the "striking, and *totally unforeseeable*, beginning of a rupture" with the existing.[119] New communists like Badiou treat the Commune as a contingent event that "brought about an immanent overturning of the laws of appearing."[120] Žižek takes left mysticism even further, seeing in the Paris Commune the manifestation of "divine violence."[121] These are simply rehearsals of the "supernatural nonsense" of "instantaneous" communism.

Whether articulated in the writings of Badiou, Žižek, Negri, or Hardt, at work here is the assumption that, as the writing group Endnotes puts it, "it is no longer possible to imagine a transition to communism on the basis of a prior victory of the working class *as* working class. To hold to councilist or Leninist conceptions of revolution now is utopian, measuring reality against mental constructs which bear no historical actuality."[122] The assumption here is that class is not a material relationship (of owning); it is a now outdated (idealist) construct. Leaving aside the idealism of notions of the common and spontaneous communism, the implication here is that, if communism is to be preceded by a socialist revolution in which the workers come to power, such a revolution is no longer possible because the working class no longer exists. Socialism has, on these terms, been made irrelevant. And in the absence of the working class, working class revolution, and socialism, communism is to be achieved as an event in which the divisions of contemporary capitalism spontaneously disappear altogether. This is the utopian desire for communism before communism. Which is to say, capitalist communism.

Kristin Ross' more recent book on the Paris Commune and its legacy, *Communal Luxury*, takes up a similar reading of the Commune, treating it more in terms of the common culture of the working class in the Commune, and aggressively opposing the arguments of Marx, Engels, and Lenin that it represented the need for the dictatorship of the proletariat (socialism) in the transition from capitalism to communism. According to Ross, "The Communal imagination operated on the preferred scale of the local autonomous unit within an internationalist horizon. It had little room for the nation, or, for that matter, for the market or the state" and was instead based on "principles of cooperation and association."[123] The "form of the Commune," Ross writes, "was less a form than a set of dismantling acts, the critique-in-act of the bureaucratic state."[124] While, in saying so, she affirms Marx's argument that "the working class cannot simply lay hold of the ready-made state machinery and wield it for its own purpose,"[125] her emphasis is on the "autonomy" of the Commune (a favorite trope of the new communists) and its refusal of the state (even as the Commune began to take on the form of the workers' state, and through it, to issue decrees and policies in the interests of working people). Significantly, one of the major concerns of Ross' book is to demonstrate the contributions of anarchists and anarchist communists, like Peter Kropotkin and Élisée Reclus, to the Paris Commune and to trace the development of their anarchist thinking before and after it.

Ross correctly observes that,

> What the commune as political and social medium offered that the factory did not was a broader social scope—one that included women, children, the peasantry, the aged and the unemployed. It comprised not merely the realm of production but both production and consumption. Differences in emphasis can be detected in different thinkers, but what the free commune entails for each is the simultaneous dissolution of Capital, State, and Nation.[126]

However, such a society cannot be achieved without workers taking control of the institutions of existing society, by seizing the means of production and the state. While the goal of communism is a society without class exploitation, so long as the relations of production and the state apparatus remains in the hands of the owners of capital, the lives of the vast majority will be limited to the poverty and alienation produced by private property. Working class control of the state is necessary to *prevent* the ruling class from exploiting others. It is not just utopian to believe capital's massive military might will not be used to crush working class efforts to bring about a more equal world, it is also symptomatic of what Lenin called left-wing "childishness"[127] that in effect sabotages the proletarian struggle. The transition from socialism to communism is necessary because, as he writes, even when the working class does come to power, "It is still necessary to suppress the bourgeoisie and crush their resistance. This was particularly necessary for the Commune; and one of the reasons of its defeat was that it did not do this with sufficient determination."[128] As he explains, "The dictatorship of the proletariat" means a most determined war waged by "the new class against its more powerful enemy, the bourgeoisie, whose power of resistance increases tenfold after its overthrow, even though overthrown in only one country. The power of the bourgeoisie rests not alone upon international capital, upon its strong international connections, but also upon the force of habit, on the force of small industry," which "daily, hourly, gives birth to capitalism and bourgeoisie, spontaneously and on a large scale."[129] As Donny Gluckstein makes clear with regard to the aftermath of the Paris Commune, the "bloody week" of the Commune's defeat "was a graphic example of a capitalist state stripped to its bare essentials—'armed bodies of men'—exterminating a threat to the system of domination and exploitation."[130] The capitalist state has proven on innumerable occasions before and since the Commune that it will not resist such carnage when the exploited work together as a movement to collectively bring about a fundamentally new world.

The writings of Badiou, Hardt, Negri, Ross, and Žižek are important because they demonstrate the way in which the new communist readings of the Commune invert the political to the ethical, proposing that radical praxis is essentially an ethical mode of being, a posthumanist ontology conceptualized as an open and indeterminate space of "life": a mode of "living" that resists what is perceived as rigid "determinism" and abstractions that explain a situation outside its "immanence." The resistance that seems to "appear" out of nowhere and just as quickly dissipate is embraced by such writers, not as a sign of proletarianization and growing class discontent—and certainly not as a sign of the need for an international working class movement and party—but as a sign of the multitude's autonomous and excessive desire to merge affectively and momentarily in common feeling, which is represented both as conceptually beyond measure and as the basis of fundamental change. In this way, Žižek also proposes that new social movements, like those in Rio de Janeiro in the 1990s, "when crowds descended from the favelas into the wealthy part of the city and started looting and burning supermarkets," are significant because they "strike[] from out of nowhere" and are "a means without an end."[131] Žižek, who suggests such struggles are important only insofar as they are void of an "end" or purpose, reads the uprising through the lens of Kantian aesthetics. But not only does this conveniently avoid the class position of those who live in the slums and erase the relation of looting and property destruction to the relations of private property that deny the world's vast majority of working people the ability to meet their needs. Žižek's reading also translates the class struggles that erupt against the ruling class into more or less aesthetic spectacles for the intellectuals of the global North.

It is by grasping society as a dialectical totality that, in his reading of the Commune, Marx shows how all social antagonisms—as between debtors and creditors, secular education against religious indoctrination, prostitutes against "the absconding men of family, religion and above all, of property,"[132] international cooperation against nationalist war—are, in the final analysis, forms of the conflict between the producing class against the appropriating class. The fact that today workers around the world are continually brought together is an effect of the ever deepening contradiction between the forces of production, which lead to more and more global social cooperation, and the social relations of production, which limit cooperative production to the narrow relations of private property and thus become fetters to the productive process. The development of productive forces, to say it again, does not in itself constitute a new common society in which the old class conflicts disappear, but instead points to the necessity of abolishing the social relations of production founded on private ownership. Revolutionary communism is "the most radical rupture with traditional property relations."[133] The transatlantic left's championing of commonism is ultimately the communism of capital that, in a militant rhetoric, limits change to change within capitalism.

BIBLIOGRAPHY

Badiou, Alain. *The Communist Hypothesis*. New York: Verso, 2010.

---. "The Idea of Communism." In *The Idea of Communism*, edited by Costas Douzinas and Slavoj Žižek, 1–14. New York: Verso, 2010.

Badiou, Alain, and Bruno Bosteels. *Philosophy for Militants*. London: Verso, 2012.

Bolt, Mikkel. *Crisis To Insurrection: Notes on the Ongoing Collapse*. Wivenhoe, UK: Minor Compositions, 2015.

Camatte, Jacques. *Capital and Community: The Results of the Immediate Process of Production and the Economic Works of Marx*. London: Unpopular Books, 1998.

Coetzee, J. M. *Life & Times of Michael K*. New York: Viking Press, 1984.

Curcio, Anna and Ceren Özselçuk. "Introduction: The Common and the Forms of the Commune." *Rethinking Marxism* 22, no. 3 (2010): 304–11.

Dean, Jodi. *The Communist Horizon*. New York: Verso, 2012.

---. "The Party and Communist Solidarity." *Rethinking Marxism: A Journal of Economics, Culture & Society* 27, no. 3 (2015): 332–42.

de Mattis, Leon. "Reflections on the Call." In *Communization and its Discontents: Contestation, Critique and Contemporary Struggles*, edited by Benjamin Noys, 61–79. New York: Minor Compositions, 2011.

---. "What is Communisation?" *SIC: International Journal for Communisation* 1, no.1 (November 2011): 11–28. Accessed January 1, 2015. https://libcom.org/book/export/html/51086.

Diskin, Jonathan. "How Subjectivity Brings Us through Class to the Community Economy."*Rethinking Marxism* 25, no. 4 (2013): 469–82.

Douzinas, Costas and Slavoj Žižek, eds. *The Idea of Communism*. New York: Verso, 2010.

Ebert, Teresa and Mas'ud Zavarzadeh. "The Digital Metaphysics of Cognitive Capitalism: Abandoning Dialectics, the North Atlantic Left Invents a Spontaneous Communism within Capitalism." *International Critical Thought* 4, no. 4 (2014): 397–417.

Endnotes. "Communization and Value-Form Theory." *Endnotes* 2 (April 2010). Accessed July 29, 2015. http://endnotes.org.uk/en/endnotes-communisation-and-value-form-theory.

---. "What are we to do?" In *Communization and its Discontents: Contestation, Critique and Contemporary Struggles*, edited by Benjamin Noys, 23–38. New York: Minor Compositions, 2011.

Engels, Frederick. Introduction to *The Civil War in France*. In *The Civil War in France and The Paris Commune*, by Karl Marx and V. I Lenin, 9–22. New York: International Publishers, 1984.

Hancox, Dan. "Why Ernesto Laclau Is the Intellectual Figurehead for Syriza and Podemos." February 9, 2015. http://www.theguardian.com/commentisfree/2015/feb/09/ernesto-laclau-intellectual-figurehead-syriza-podemos.

Hardt, Michael. "The Common in Communism." *Rethinking Marxism* 22, no. 3 (2010): 346–56.

Hardt, Michael and Antonio Negri. *Commonwealth*. Cambridge, MA: Belknap Press, 2009.

---. *Empire*. Cambridge: Harvard University Press, 2000.

---. *Multitude: War and Democracy in the Age of Empire*. New York: Penguin Press, 2004.

Healy, Stephen. "Communism as a Mode of Life." *Rethinking Marxism: A Journal of Economics, Culture & Society* 27, no. 3 (2015): 343–56.

Gibson-Graham, J. K. *The End of Capitalism (As We Knew It): A Feminist Critique of Political Economy*. Minneapolis: University Minnesota Press, 2006.

Gluckstein, Donny. *The Paris Commune: A Revolution in Democracy*. Chicago, IL: Haymarket, 2011.

Guattari, Félix, and Antonio Negri. *Communists Like Us: New Spaces of Liberty, New Lines of Alliance*. New York: Semiotext(e), 1990.

Kantor, Jodi and David Streitfeld, "Inside Amazon: Wrestling Big Ideas in a Bruising Workplace." *New York Times*, August 15 (2015). Accessed August 15, 2015. http://www.nytimes.com/2015/08/16/technology/inside-amazon-wrestling-big-ideas-in-a-bruising-workplace.html.

Kay, Joseph. "Keep on Smiling," February 8, 2007 (19:20), *Aufheben* 14, http://libcom.org/library/aufheben/aufheben-14-2006/keep-on-smiling-questions-on-immaterial-labour.

Klein, Naomi. *This Changes Everything: Capitalism Vs. the Climate*. New York: Simon & Schuster, 2014.

Laclau, Ernesto and Chantal Mouffe. *New Reflections on the Revolution or Our Time*. New York: Verso, 1990.

Lazzarato, Maurizio. "From Capital-Labour to Capital-Life." *Ephemera: Theory and Politics in Organization* 4, no. 3 (2004): 187–208. Accessed July 22, 2015. http://www.ephemerajournal.org/contribution/capital-labour-capital-life.

Lenin, V. I. *"Left-Wing" Communism: An Infantile Disorder*. Detroit: The Marxian Educational Society, 1921.

---. "Marx's Assessment of the Commune." In *The Civil War in France and The Paris Commune*, by Karl Marx and V. I Lenin, 91–95. New York: International Publishers. 1984.

---. *State and Revolution*. In *The Lenin Anthology*, edited by Robert Tucker, 311–98. New York: W. W. Norton, 1975.

---. *What Is To Be Done? Burning Questions of Our Movement*. New York: International Publishers, 1969.

Luisetti, Federico. "The Nature of the Common." *Rethinking Marxism* 22, no. 3 (2010): 434–36.

Mackay, Robin and Armen Avanessian, eds. *The Accelerate Reader*. Falmouth, UK: Urbanomic Media 2014.

Mackay, Robin and Armen Avanessian, eds. Introduction to *The Accelerate Reader*, 1–46. Falmouth, UK: Urbanomic Media, 2014.

Marx, Karl. *Capital*, vol. 1. Translated by Ben Fowkes. New York: Penguin, 1976.

---. *The Civil War in France*. In *The Civil War in France and The Paris Commune*, by Karl Marx and V. I Lenin, 23–82. New York: International Publishers. 1984.

---. *Grundrisse*. Translated by Martin Nicolaus. New York: Penguin, 1976.

---. "Outlines of *The Civil War in France*." In *On the Paris Commune*, by Karl Marx and Frederick Engels, 102–221. Moscow: Progress Publishers, 1971.

---. "Resolutions of the Meeting Held to Celebrate the Anniversary of the Paris Commune." In *On the Paris Commune*, 269. Moscow: Progress Publishers, 1971.

Marx, Karl and Frederick Engels. *The German Ideology:* In *Collected Works*, vol. 5. New York: International Publishers, 1976: 19–539.

---. *Manifesto of the Communist Party*. In *Marx-Engels Collected Works*, vol. 6. New York: International Publishers, 1976: 477–519.

---. *On the Paris Commune*. Moscow: Progress Publishers, 1971.

Moulier Boutang, Yann, and Ed Emery. *Cognitive Capitalism*. Cambridge, UK: Polity Press, 2011.

Negri, Antonio. *Goodbye Mr. Socialism*. New York: Seven Stories Press, 2008.

---. "The Labor of the Multitude and the Fabric of Biopolitics." *Mediations: Journal of the Marxist Literary Group* 23, no. 2 (2008): 9–24. http://www.mediationsjournal.org/articles/the-labor-of-the-multitude-and-the-fabric-of-biopolitics.

---. "On Commonwealth." Translated by Arianna Bove. *Generation Online* (October 2009). Accessed July 29, 2015. http://generation-online.org/p/fp_negri22.htm.

---. "Some Reflections on the #Accelerate Manifesto." In *The Accelerate Reader*, edited by Robin Mackay and Armen Avanessian, 363–78. Falmouth, UK: Urbanomic Media, 2014.

Noys, Benjamin, ed. *Communization and its Discontents: Contestation, Critique and Contemporary Struggles*. New York: Minor Compositions, 2011.

Out of the Woods Collective. "Disaster Communism." Accessed August 21, 2015. http://libcom.org/blog/disaster-communism-part-3-logistics-repurposing-bricolage-22052014.

Reed, Patricia. "Seven Prescriptions for Accelerationism." In *The Accelerate Reader*, edited by Robin Mackay and Armen Avanessian, 521–36. Falmouth, UK: Urbanomic Media, 2014.

Roggero, Gigi. "Five Theses." *Rethinking Marxism: A Journal of Economics, Culture & Society* 22, no. 3 (2010): 357–73.

Ross, Kristin. *Communal Luxury: The Political Imaginary of the Paris Commune*. New York: Verso, 2015.

Streitfeld, David. "Data-Crunching is Coming to Help your Boss Manage Your Time." *New York Times*, August 17, 2015. Accessed Aug. 20, 2015. http://www.nytimes.com/2015/08/18/technology/data-crunching-is-coming-to-help-your-boss-manage-your-time.html.

Théorie Communiste, "Communization in the Present Tense." In *Communization and its Discontents: Contestation, Critique and Contemporary Struggles*, edited by Benjamin Noys, 41–58. New York: Minor Compositions, 2011.

Terranova, Tiziana. "Red Stack Attack!" In *The Accelerate Reader*, edited by Robin Mackay and Armen Avanessian, 379–99. Falmouth, UK: Urbanomic Media, 2014.

Torrant, Julie. *The Material Family*. Rotterdam, The Netherlands: Sense Publishers, 2011.

Varoufakis, Yanis. "How I Became an Erratic Marxist." *The Guardian*. February 18, 2015. Accessed October 7, 2015. http://www.theguardian.com/news/2015/feb/18/yanis-varoufakis-how-i-became-an-erratic-marxist.

Virno, Paolo. *A Grammar of the Multitude: For an Analysis of Contemporary Forms of Life*. New York: Semiotext(e), 2004.

Wall, Derek. *The Common in History: Culture, Conflict and Ecology*. Cambridge, MA: MIT Press, 2014.

Williams, Alex and Nick Srnicek, "#Accelerate Manifesto: For an Accelerationist Politics."*Critical Legal Thinking*. May 14, 2013. Accessed May 2, 2015. http://criticallegalthinking.com/2013/05/14/accelerate-manifesto-for-an-accelerationist-politics/.

Žižek, Slavoj. "How to Begin from the Beginning." In *The Idea of Communism*, edited by Costas Douzinas and Slavoj Žižek, 209–26. New York: Verso, 2010.

---. *In Defense of Lost Causes*. London: Verso, 2008.

NOTES

1. Karl Marx and Frederick Engels, *Manifesto of the Communist Party*, in *Marx-Engels Collected Works*, vol. 6 (New York: International Publishers, 1976), 516.

2. Michael Hardt and Antonio Negri, *Commonwealth* (Cambridge, MA: Belknap Press, 2009), viii.

3. Stephen Healy, "Communism as a Mode of Life," *Rethinking Marxism: A Journal of Economics, Culture & Society* 27, no. 3 (2015): 343, 345.

4. J. M. Coetzee, *Life & Times of Michael K* (New York: Viking Press, 1984), 99.

5. . Michael Hardt and Antonio Negri, *Empire* (Cambridge, Mass: Harvard University Press, 2000), 203.

6. Hardt and Negri, *Empire*, 204.

7. Ibid., 195.

8. Ibid., xii.

9. Ibid., 204.

10. Ibid., 204.

11. For a discussion of the common in free-market proponents like Wales and Ostrom, see Derek Wall, *The Common in History: Culture, Conflict and Ecology* (12; 72).

12. Karl Marx, *The Civil War in France*, in *The Civil War in France and The Paris Commune*, by Karl Marx and V. I. Lenin (New York: International Publishers, 1984), 60–61.
13. Marx, Karl and Frederick Engels, *The German Ideology*, in *Marx-Engels Collected Works*, vol. 5 (New York: International Publishers, 1976), 50–53.
14. Marx and Engels. *The German Ideology*, 51.
15. Ibid., 52.
16. Marx, *Civil War in France*, 81.
17. Karl Marx, "First Outline of *The Civil War in France*," in *On the Paris Commune* by Karl Marx and Frederick Engels (Moscow: Progress Publishers, 1971), 152.
18. Hardt and Negri, *Empire*, 31.
19. Michael Hardt and Antonio Negri, *Multitude: War and Democracy in the Age of Empire* (New York: Penguin Press, 2004), 83.
20. Hardt and Negri, *Multitude*, 69.
21. Ibid., 91–93.
22. V. I. Lenin, *What Is To Be Done? Burning Questions of Our Movement* (New York: International Publishers, 1969), 25 (my emphasis).
23. Antonio Negri, "On Commonwealth," trans. Arianna Bove, *Generation Online* (October 2009), accessed July 29, 2015. http://generation-online.org/p/fp_negri22.htm.
24. Negri, "On Commonwealth," n.p.
25. See Federico Luisetti's argument that "the natural outside of the political subject" is what is "missed" by "post-Marxist political thinking" on the common, in "The Nature of the Common," *Rethinking Marxism* 22, no. 3 (2010): 435.
26. Paolo Virno, *A Grammar of the Multitude: For an Analysis of Contemporary Forms of Life* (New York: Semiotext(e), 2004),110.
27. Ibid., 110.
28. Yann Moulier Boutang, *Cognitive Capitalism* (Cambridge, UK: Polity Press, 2011), 9.
29. Ibid., 6.
30. Ibid., 57.
31. Hardt and Negri, *Empire*, 302.
32. Maurizio Lazzarato, "From Capital-Labour to Capital-Life," *Ephemera: Theory and Politics in Organization* 4, no. 3 (2004), 197, accessed July 22, 2015, http://www.ephemerajournal.org/contribution/capital-labour-capital-life.
33. Félix Guattari and Antonio Negri, *Communists Like Us: New Spaces of Liberty, New Lines of Alliance* (New York: Semiotext(e), 1990), 22.
34. Ibid., 22.
35. Michael Hardt and Antonio Negri, *Commonwealth* (Cambridge, MA: Belknap Press, 2009), x.
36. Hardt and Negri, *Empire*, 365.
37. Antonio Negri, "The Labor of the Multitude and the Fabric of Biopolitics," *Mediations: Journal of the Marxist Literary Group*, 23, no. 2 (2008), 20, accessed January 1, 2016, http://www.mediationsjournal.org/articles/the-labor-of-the-multitude-and-the-fabric-of-biopolitics.
38. Karl Marx, *Capital*, vol. 1, trans. Ben Fowkes (New York: Penguin, 1976), 142.
39. It is not a coincidence that the defense of "immaterial labor" as the basis of creativity also involves the appeal to the "matter" of bodies ("viscera") and affect: through the appeal to affect, biopolitics, and life, the materialist analysis of the place of individuals, ideas, and bodies in the social relations of production is displaced with matterism, the idealist concern with bodies and affects as self-generating and autonomous sites of resistance. Through such displacements, which are often advanced in the "more materialist" name of the "matter" of the "concrete" as opposed to the abstract, society as an interconnected totality disappears, backgrounded by nuanced appeal to "differences," which are given high philosophical (and increasingly "ontological") significance as signs of the individual's "singularity" within the "common."
40. Lazzarato, "From Capital-Labour to Capital-Life," 203-4.

41. Michael Hardt, "The Common in Communism," *Rethinking Marxism* 22, no. 3 (2010), 349.
42. Marx, *Capital*, 272–73.
43. Hardt and Negri, *Multitude*, 103.
44. Ibid., 104.
45. Anna Curcio and Ceren Özselçuk, "The Common and the Forms of the Commune," *Rethinking Marxism* 22, no. 3 (2010): 309.
46. See, for instance, Jonathan Diskin's writing on the work of J. K. Gibson-Graham, especially his emphasis on the "critique of capitalism and capitalocentrism," or those (Marxist) theories that treat "capitalism as the figure for a closed and total representation of the economy" and which consequently "conspire to make us feel small, our efforts to live beyond capitalism marginal and weak" ("How Subjectivity Brings Us through Class to the Community Economy," *Rethinking Marxism* 25, no. 4 [2013]: 470; 478; 471).
47. Curcio and Özselçuk, "The Common and the Forms of the Commune," 308.
48. Hardt and Negri, *Commonwealth*, ix.
49. Tiziana Terranova, "Red Stack Attack!" in *The Accelerate Reader*, ed. Robin Mackay and Armen Avanessian (Falmouth, UK: Urbanomic Media, 2014) 382; 383.
50. Ibid., 386.
51. Jodi Dean, "The Party and Communist Solidarity," *Rethinking Marxism: A Journal of Economics, Culture & Society* 27, no. 3 (2015), 332–33.
52. Ibid., 335.
53. Ibid., 335.
54. Out of the Woods, "Disaster Communism," *libcom.org* (blog), May 22, 2014 (9:57am), http://libcom.org/blog/disaster-communism-part-3-logistics-repurposing-bricolage-22052014.
55. Naomi Klein, *This Changes Everything: Capitalism vs. the Climate* (New York: Simon & Schuster, 2014), 352.
56. Marx, *Capital*, 439.
57. Ibid., 449.
58. Ibid., 438.
59. Karl Marx, *Grundrisse*, trans. Martin Nicolaus (New York: Penguin, 1976), 704.
60. Marx and Engels, *Manifesto*, 504.
61. Endnotes, "Communization and Value-Form Theory," *Endnotes 2* (April 2010), accessed July 29, 2015, http://endnotes.org.uk/en/endnotes-communisation-and-value-form-theory.
62. Robin Mackay and Armen Avanessian, introduction to *The Accelerate Reader* (Falmouth, UK: Urbanomic Media, 2014), 15.
63. Mackay and Avanessian, Introduction, 39.
64. See, for instance, Alex Williams and Nick Srnicek's "#Accelerate Manifesto: For an Accelerationist Politics" and Mackay and Avanessian's introduction to *The Accelerate Reader* for a discussion of some of the immanent differences. For a critique of accelerationism, see my forthcoming *Materialism and Resistance*.
65. Marx, *Capital*, 1025–34.
66. Ibid., 1024.
67. Ibid., 451.
68. For a sustained materialist analysis of the family, see Julie Torrant's *The Material Family*.
69. David Streitfeld, "Data-Crunching is Coming to Help your Boss Manage Your Time," *New York Times* August 17, 2015, accessed August 20, 2015, http://www.nytimes.com/2015/08/18/technology/data-crunching-is-coming-to-help-your-boss-manage-your-time.html.
70. Jodi Kantor and David Streitfeld, "Inside Amazon: Wrestling Big Ideas in a Bruising Workplace," *New York Times* August 15, 2015, accessed August 15, 2015, http://www.nytimes.com/2015/08/16/technology/inside-amazon-wrestling-big-ideas-in-a-bruising-workplace.html.

71. Kantor and Streitfeld, "Inside Amazon."
72. Mikkel Bolt, *Crisis to Insurrection: Notes on the Ongoing Collapse* (Wivenhoe: Minor Compositions, 2015), 8.
73. Teresa Ebert and Mas'ud Zavarzadeh, "The Digital Metaphysics of Cognitive Capitalism: Abandoning Dialectics, the North Atlantic Left Invents a Spontaneous Communism within Capitalism," *International Critical Thought* 4, no. 4 (2014): 408.
74. Ibid., 408.
75. Ibid., 408.
76. Antonio Negri, "Some Reflections on the #Accelerate Manifesto," in *Accelerate Reader*, ed. Robin Mackay and Armen Avanessian (Falmouth, UK: Urbanomic Media, 2014), 367–68.
77. Joseph Kay, "Keep on Smiling," *libcom.org*, February 8, 2007 (19:20), http://libcom.org/library/aufheben/aufheben-14-2006/keep-on-smiling-questions-on-immaterial-labour.
78. Gigi Roggero, "Five Theses." *Rethinking Marxism: A Journal of Economics, Culture & Society* 22, no. 3 (2010): 364 .
79. Leon de Mattis, "What is Communisation?" *SIC: International Journal for Communisation* 1, no. 1 (Nov. 16, 2011), 24, accessed January 1, 2016, https://libcom.org/book/export/html/51086.
80. Ibid., 23.
81. Ibid., 23.
82. Endnotes, "Communization and Value-Form Theory."
83. Ibid.
84. de Mattis, "What is Communisation?", 24.
85. Ibid., 26.
86. Jacques Camatte, *Capital and Community: The Results of the Immediate Process of Production and the Economic Works of Marx* (London: Unpopular Books, 1998), 148, n.19.
87. V. I. Lenin, *"Left-Wing" Communism: An Infantile Disorder* (Detroit: The Marxian Educational Society, 1921), 37; 61.
88. Ibid., 27–28.
89. Ibid., 28.
90. Ibid., 26.
91. Ibid., 38–39.
92. Curcio and Özselçuk, Introduction, 308.
93. Ibid., 308.
94. Patricia Reed, "Seven Prescriptions for Accelerationism," in *The Accelerate Reader*, ed. Robin Mackay and Armen Avanessian (Falmouth, UK: Urbanomic Media, 2014), 534.
95. Marx and Engels, *Manifesto*, 494.
96. V. I. Lenin, "Marx's Assessment of the Commune," in *The Civil War in France and The Paris Commune*, by Karl Marx and V. I. Lenin (New York: International Publishers. 1984), 92.
97. Marx, *The Civil War in France*, 60.
98. Ibid., 61, 62.
99. Cited in "Resolutions of the Meeting Held to Celebrate the Anniversary of the Paris Commune," *On the Paris Commune* (Moscow: Progress Publishers, 1971), 269.
100. Marx, *The Civil War in France*, 51.
101. Donny Gluckstein, *The Paris Commune: A Revolution in Democracy* (Chicago, IL: Haymarket, 2011), 149.
102. Ibid., 162.
103. Frederick Engels, introduction to *The Civil War in France*, in *The Civil War in France and The Paris Commune*, by Karl Marx and V. I. Lenin (New York: International Publishers, 1984), 18.
104. Engels, Introduction, 18.
105. Ibid., 22.
106. Marx, *The Civil War in France*, 81.

107. Marx, "Outlines of The Civil War in France," in *On the Paris Commune*, by Karl Marx and Frederick Engels (Moscow: Progress Publishers, 1971), 152.
108. Michael Hardt, "The Common in Communism," *Rethinking Marxism* 22, no. 3 (2010): 346.
109. Hardt and Negri, *Multitude*, 69.
110. To this point, in contrast to proponents of immediate "communinization" (like de Mattis), Lenin emphasizes that "We are not utopians, we do not 'dream' of dispensing at once with all administration, with all subordination. These anarchist dreams [. . .] serve only to postpone the socialist revolution." *State and Revolution*, in *The Lenin Anthology*, ed. Robert Tucker (New York: Norton, 1975), 344.
111. Dan Hancox, "Why Ernesto Laclau Is the Intellectual Figurehead for Syriza and Podemos," *The Guardian* (February 9, 2015), http://www.theguardian.com/commentisfree/2015/feb/09/ernesto-laclau-intellectual-figurehead-syriza-podemos.
112. Ernesto Laclau and Chantal Mouffe, *New Reflections on the Revolution or Our Time* (New York: Verso, 1990), 89–92.
113. Yanis Varoufakis, "How I Became an Erratic Marxist," *The Guardian* (February 18, 2015), http://www.theguardian.com/news/2015/feb/18/yanis-varoufakis-how-i-became-an-erratic-marxist.
114. Marx and Engels, *Manifesto*, 514.
115. Alain Badiou, *The Communist Hypothesis* (New York: Verso, 2010), 222.
116. Ibid., 200.
117. Ibid., 208.
118. Ibid., 178.
119. Ibid., 206; my emphasis.
120. Ibid., 206.
121. Slavoj Žižek, *In Defense of Lost Causes* (London: Verso, 2008), 161.
122. Endnotes, "What are we to do?", in *Communization and its Discontents: Contestation, Critique and Contemporary Struggles*, ed. Benjamin Noys (New York: Minor Compositions, 2011), 27.
123. Kristin Ross, *Communal Luxury: The Political Imaginary of the Paris Commune* (New York: Verso, 2015), 5.
124. Ross, *Communal Luxury*, 79.
125. Ibid., 78.
126. Ibid., 111–12.
127. Lenin, "'Left-Wing' Communism," 61.
128. V. I. Lenin, *State and Revolution*, in *The Lenin Anthology*, ed. Robert Tucker, (New York: W. W. Norton, 1975), 340.
129. Lenin, "'Left-Wing' Communism," 16.
130. Gluckstein, *The Paris Commune*, 159.
131. Žižek, *In Defense of Lost Causes*,162.
132. Marx, *The Civil War in France*, 68.
133. Marx and Engels, *Manifesto*, 504.

SEVEN

Posthumanist Metaphysics and the Necessity of Dialectics

Robert Faivre

Posthumanism is a class metaphysics through which an array of emergent crisis issues, such as climate change, extremes of poverty and hunger, and epidemics of preventable illness and death, can be acknowledged as problems and addressed through temporary and local reforms which target the effects but leave the root causes invisible and untouched. Through figures and logics of the human in crisis, posthumanism represents the structural contradictions of capitalism and the historical relations of class as existential and ontological conditions of life as such, mystifying these as beyond knowledge and social decision. That is, posthumanism, as an update of bourgeois ideology, is a theory for managing the contradictions and crises of global capital; it is not a theory for change but a theory for accommodation to what is taken to be unchangeable or inevitable on a global scale. In its theoretical articulations and in its literary-cultural dramatizations—such as the emergent genre of climate fiction—posthumanist metaphysics plays out a self-contradictory account of social structure and agency at a time of crisis and global change. In contrast to the discourses of posthumanism and their theoretical basis in contemporary metaphysics, dialectics is the explanation of the necessity of contradictions and historical materialism is the theory for ending the interminable crises of capitalism and establishing new relations which do not require metaphysics.

Metaphysics, which renders reality in excess of human understanding, has long been opposed to by dialectics, which in the tradition of Marx is the science of social transformation: a materialist explanation of how and why both nature and the historical relations of human society develop and in turn determine and structure ideas. In its posthumanist iteration, metaphysics produces a cultural imaginary in which social reality and nature are comprised of a chaotic series of accidents and events, which are ultimately treated as products of (human) nature and thus placed beyond understanding and social change. The deep interconnections of social and natural relations are, through posthumanist metaphysics, disconnected from each other and above all from the mode of production that structures humans' relations with each other and with the natural world. The ideas of posthumanism update the metaphysics by which bourgeois ideology naturalizes the class interests of capital in an era of developing crises. In the guise of claiming to address the problems facing not only "humanity" but all life on the planet, the metaphysics of posthumanism in fact advance the interests of the dominant class in maintaining the basic relations of class exploitation.

Posthumanism as part of the cultural commonsense of capitalism achieves its effects primarily by denying the explanatory value of dialectics. Rosi Braidotti, for example, articulates a common rejection of dialectics as "anthropocentric, gendered, racialized,"[1] and "violent."[2] She writes, "The dialectics of otherness is the inner engine of humanist Man's power, who assigns difference on a hierarchical scale as a tool of governance."[3] As an egalitarian alternative (what Braidotti terms a "*zoē*-egalitarian" alternative, referring to her vitalist theory of life-force) to humanist "dominance and exclusion,"[4] what posthumanism advocates is an account of difference without the "violence" of dialectical "oppositional thinking" which reduces difference to a power relation through "instrumental, narrow logic."[5] Braidotti's metaphysical conflation of differences of identity with hierarchies of value, which is then projected onto materialist theory as a rationale for rejecting dialectics, is an ideological manoeuver for avoiding contradiction. She denies the objective material contradictions of exploitation and inequality and represents differences as outside of the material contradictions of capitalism. Materialist theory and its practice of dialectics explains social differences and values in terms of objective material contradictions which require transformation to be resolved. The irony—and an ideological marker of posthumanism as metaphysics—is that posthumanism rejects materialist dialectics precisely because materialism explains and enables the resolution of the contradiction which produces posthumanism as an ideological discourse. Rather than being the means by which humanism imposes order on conditions of freedom, which is how posthumanism represents materialist dialectics as part of the tradition of racist, sexist, and anthropocentric violence, dialectics is developed by materialists as the explanation of the class contradiction at the base of violence and discrimination based on cultural difference. But because materialist dialectics advances from the recognition of the priority of the class contradiction, ideological defenses of the social relations of class develop not only to naturalize the class contradiction but indeed to recast contradictions relations of difference beyond dialectical opposition. Thus, if the very real contradictions to which posthumanism is an ideological response are to be understood, explained, and transformed, then dialectics as the theory of contradiction must be advanced as integral to the collective building of a truly posthumanist world: a post-capitalist society in which the contradiction between exploiter and exploited has been overturned.

As John Rees rigorously outlines in *The Algebra of Revolution*, dialectics emerges historically as a materialist correction of idealist and speculative modes of philosophical and scientific theory. That is, dialectics emerges out of the real contradictions which metaphysics as ideology cannot resolve but can only reproduce and recontain. Materialist dialectics, Rees explains, builds its account of reality and change through the use of a range of key concepts, including primarily among them: totality, contradiction, and change. Totality is the concept that what we confront in thought and practice is an interrelated whole, differentiated but interrelated, characterized by unity of the diverse. That the differentiated aspects of the whole are interrelated is conceptualized as mediation, the way in which each part expresses the overall character of the whole in its relations, and that the whole is mediated by the interrelation of the parts. Not only are the parts diverse, thus potentially contradictory, but each element of the structure is not an isolated thing-in-itself but is coherent only in relation to other elements and indeed is self-contradictory in its development, its becoming other; in fact, contradiction is the form of the primary relations of totality, in that contradiction is the root source of all change within totality, and the social contradiction of class is the basis of crisis and change in capitalism. In this sense, as Rees writes, what makes materialist dialectics "critical and revolutionary"[6] in its essence is its ability to account for and explain the dynamic interrelations between the abstract and the concrete, the local and the global, and the existing and the possible, *not* as the effects of ideas, but as the movement of the social totality. Thus, in contrast to the idealism of the Hegelian dialectic, according to which "the real world is only the external, phenomenal form of 'the idea,'" materialist dialectics treats the "ideal" as "nothing else than the material world reflected by the human mind, and translated into forms of thought."[7] Rees states, "we are dealing with real social processes and so the dialectic has to be able to show how the central contradiction of capitalist society is expressed, often in very different forms, in all of the economic, cultural, political, ideological, and legal aspects of society."[8] Materialist dialectics develops out of Marx's discovery that the core contradiction of capitalism is the exploitative relation of class at the site of production, and that this basic class antagonism—between the owning class and the laboring class—is the definitive mediation throughout the social totality and the secondary relations therein. Indeed, nature, the other source of value besides human labor-power, is also a theatre of class antagonism between the bourgeoisie's incessant drive for profit and the proletariat's struggle for not only survival but social transformation. Dialectics, through the development of the materialist concepts and logics whereby nature and society can be explained, is everywhere opposed by the cultural commonsense of capitalism, which marginalizes dialectics and takes up the ideological metaphysics of anti-dialectics because dialectics is the theory of contradiction and therefore needed to critique class contradiction.

The question of fossil fuels offers a useful example to demonstrate the contradictory relation of dialectics and metaphysics as class theories. Since the Industrial Revolution, a few countries and corporations together generate more greenhouse gases than the majority of the global population, and yet the forecasted consequences will likely affect most those who have had the least role in causing the problem. Not only is this a contradiction, but it is an expression of the basic class contradiction of property relations and thus decision over the priorities of production. For, in both the regions in which most greenhouse gases are produced and the regions most affected by their destructive effects, it is not the working class but the capitalist monopolization of social resources for the appropriation of surplus value that is responsible for the ever-widening carbon footprint. As a result, it is in the class interest of the proletariat in the global north and the global south to join forces and fight transnational capital's subjugation of the world's population and the earth to the priorities of profit. This is another way of saying that it is not "the human race" as a whole which has decided to rely on fossil fuels for energy, just as it is not the population as a whole which decides, for example, the location and purpose of manufacturing plants, gas refineries, or transportation infrastructure. For example, as Malm and Hornborg explain, not only was the decision to use steam engines in the early nineteenth century not made by the whole of humanity nor by popular vote within Britain, but it was made by a small "clique of British men" who were among "the owners of the means of production" in one part of the world.[9] Such decisions are made by and in the interests of one side of the class contradiction: the class which owns and controls the means of the production, the bourgeois or ruling class, which has the strongest influence over social decisions and priorities, directly and indirectly. Thus, the developing crisis of climate change is a threat to the current status quo—the contradictory property relations which allow the capitalist class to profit in production both from its exploitation of the labor-power of the working class and its command and control over the use of natural resources. But it is a threat which appears in the commonsense of capital as a crisis to be managed, a problem to be solved by making capitalism "green" or "sustainable": essentially a new mode of consumption. Because the crisis is not just an appearance of a conflict, but is the appearance of the class contradiction at the base of capitalism, the capitalist class marshals accounts and theories for reform and banishes to the margins transformative knowledge of crisis and contradiction. For instance, when climate scientist Kevin Anderson states that "As we massage the assumptions of our analysis to fit within today's political and economic hegemony, so we do society a grave disservice—one for which the repercussions will be irreversible," he hints at the way science is compromised—at great social cost—when it is made subject to the interests of the ruling class. The role of posthumanist metaphysics as a bourgeois theory is to re-contain potential transformative change and limit it to immanent change within capitalism. In this context, the claim that climate change can be addressed through a different mode of consumption cuts off consumption from the capitalist mode of production, and treats the differences of consuming habits worldwide as separate from class and the relations of imperialism. This metaphysical reading of environmental crisis addresses phenomena one-sidedly, at the level of their surface effects and not as part of an interconnected totality of capitalism. As a result, it benefits capitalists' interest in making only minor changes within capitalism, and obscures the objective reality that only through a fundamental transformation of the relations of production will humanity have any hope of changing the consumption patterns that lead to the kind of climate change associated with the Anthropocene. Materialist dialectics is the science that defends the proletarian side of the class contradiction; its goal is to theorize the development not just of change within capitalism but the conditions that lead to its overthrow and to the emergence of new relations without class.

Because the posthumanist commonsense refers to contradiction but in a way that puts it in ideological suspension as singularities outside any structural relations, indeed outside totality, there is an urgent class need for dialectical re-reading the texts of posthumanism, a dialectical critique which makes evident the historical dynamics of causality and change, and enables revolutionary knowledge, decision-making, and action. In other words, posthumanism must be read for how its own immanent contradictions—how what it proposes fails to resolve the problem of class which is at the base of issues such as climate change, energy use, population crises, and war—are expressions of the class contradiction at the base of capitalism. Further, based on such reading and analysis, materialist dialectics explains capitalism and its current crises relative to the necessity of transformation and the folly of reform.

Ian McEwan's 2010 novel of climate change, *Solar*, exemplifies the anti-dialectical ideo-logic of posthumanism today. The novel is both a parody of humanism and a posthumanist argument identifying a problem facing "humanity" and dramatizing human failure as a species trait. It thus is instructive to read the novel for its irony—that is, for how it defines the problem as solvable and yet for how it undermines solutions with assumptions about human fallibility as a fundamental aspect of being human. This irony is not only an expression of the ideological work which the novel does, but also reflects the historical contradiction of which post-humanism is a mediation—that is, the conflict between climate change and other crises as social problems that must be addressed and the economic necessities and social priorities of the capitalist class which only address such problems insofar as they impact profit and property relations. McEwan does not address the contradictions in these terms, but rather emphasizes social problems as matters of "human nature."

Discussing the novel in an online interview with the environmentalist group Friends of the Earth, McEwan raises the "particular problems for our nature" posed by global warming and climate change. Humanity as a whole must engage in "a scale of long-term thinking that lies outside our biology," which means thinking in order to "do favors for people we'll never meet" rather than thinking in terms of "trade" and "reciprocal acts."[10] Catastrophic global climate change, he states, is an "intractable subject" for a novelist without specialized scientific knowledge, one which carries a "moral load" making the novel and its reading "uncomfortable" but necessary because "we need a new energy source for civilization" and must "stop burning fossil fuels."[11] McEwan regards this to be a "comic" situation, and thus prime material for a novel. The contradiction—or what McEwan regards as a paradox—of the habituations of "human nature" and the actions shown to be necessary by climate science has literary and social import because human society is in a situation "uniquely difficult" to change and yet, McEwan posits, there is a chance that "our cleverness might win through."[12]

To illustrate this view of the problem, McEwan refers to a scene from an arctic interdisciplinary research vessel on which he travelled, a scene dramatized in the novel. The trip was an arts and science encounter in the Arctic, in which scientists and artists engaged in projects and discussions about climate change. McEwan characterizes the "boot room," the transition space between aboard the ship and the arctic landscape, wherein people changed gear, as a space

> slowly descending into the most extraordinary chaos. The discrepancy, the gap that opened up between what we wanted to do to the whole earth below us and the way we couldn't manage this room of 8 by 10 next door, suggested to me that really the way into this subject was about human nature. This really posed such a problem for human nature.[13]

Human nature, here, is an abstraction drawn out of the scene of the increasingly disorderly boot room, where cold and tired people, who nonetheless have the assurance of warmth, sustenance, and company, seemingly cannot keep their gear and clothing orderly, in contrast to the larger necessity to address the global problem of climate change, which the various fields of science make known to us and which the various fields of the arts and humanities can explore as problems. The disorderly coatroom can be explained, according to McEwan, only—and most "profoundly"—in terms of "human nature."

In the novel's dramatization of what he calls the "paradoxical" situation facing humanity today, McEwan emphasizes the contradiction between what needs to be done on a global level (which for him means a change in "fossil fuels," or a change in consumption) and what "human nature" and its habituations will allow. While each side of *Solar*'s contradiction—specifically the opposition of necessity and nature—is posited by McEwan as realistic, on inspection each is actually an ideological construct, one conditioning the other. The idea that what human society needs to do to avert crisis is to change consumption patterns, especially energy sources, while leaving aside any critical inquiry into production priorities, primarily the priority of production for private profit over freedom from need, is a ruling class idea, redirecting the focus away from the structural cause of the problem in the class relations at the site of production and onto consumption as a space of agency and ethical choice. McEwan's reading is largely informed by cutting off the problems of human reason from material conditions, rather than showing how the development of knowledge and science are determined by social relations. For instance, McEwan formulates the main opposition as an ideological one, one of a transhistorical and split "human nature" in which human knowledge and achievement (what McEwan calls "our cleverness") is in perpetual recovery from the repetition of seemingly inevitable human failures.[14] Not only is the argument that the problem is "human nature" a humanist explanation, but posthumanism does not actually break with this humanist logic but is the dissatisfaction of humanism with itself. Thus, posthumanism, as in McEwan's circular etiology of a flawed "human nature," does not resolve the limits of humanism but recycles them in other language.

In these and other such comments, McEwan sounds a central theme of posthumanism, especially in its notion of the Anthropocene, according to which fossil fuel use and population growth are thought to have made humans as a species into "a geological agent on the planet."[15] As Dipesh Chakrabarty frames it, the human species has "tumbled" by accident into a higher order of agency that is no longer only biological; rather it is a geological, planetary agency.[16] For Chakrabarty, as for McEwan, this increased agency exceeds rational understanding and attempts at social management of nature. Indeed, Chakrabarty argues that "Politics has never been based on reason alone. And politics in the age of the masses and in a world already complicated by sharp inequalities between and inside nations is something no one can control."[17] While Chakrabarty marks social inequalities, he marks these as beyond human control, as if they were facts and forces of nature. But inequality is produced by the exploitative relation of class; it is an effect of the monopolization of social resources by the few. And, of course, reason is not practiced in an analytical vacuum, separate from historical developments; rather, politics and science and other fora for discourse are theatres in which reason—ideological and counter-ideological, metaphysical and dialectical—is advanced as a form of class agency. To assert that "no one can control" the chaos of political struggles is to naturalize and obscure from view the class contradiction which shapes all politics.

Chakrabarty's reading of reason, like McEwan's analogy of the boot room, is underwritten by the assumption that planning and enacting a rational or scientific plan for averting disaster is impossible. Such impossibilities, for McEwan and Chakrabarty, are naturalized as facts of "human nature" outside history. However, humans develop their abilities not in transhistorical nature but in historical social relations. Thus, the posthumanist notion of the Anthropocene as the era of "anthropogenic" climate change, wherein humans as a *species* (by virtue of their nature) have brought about climate crisis, is an ideological notion that puts out of view the role of class relations in climate change, and advances the notion of an undifferentiated humanity, or a false universal. As Andreas Malm explains, "Humanity [. . .] is far too slender an abstraction to carry the burden of culpability, and thus that "Ours is the geological epoch not of humanity, but of capital."[18] In other words, the systematic use of fossil fuels has come about under particular historical conditions which are obscured by transhistorical appeals to human nature. Further, in their critique of Chakrabarty, Malm and Hornborg argue that because so-called "anthropogenic" climate change has as its cause not the impact of all humanity but rather the priorities and decisions of the capitalist class, it should be regarded as "sociogenic" climate change, so as not to inaccurately and ideologically attribute the effects of the actions of one class to the whole of the species.[19] A concept such as "sociogenic climate change" is dialectical to the extent that it addresses the way that social (class) relations shape the natural environment and shows that only through social transformation can the natural crises facing humanity be systematically resolved. It opens up an explanatory grasp of the root contradiction of the problem, while metaphysical abstractions such as "anthropogenic climate change" and "Anthropocene" are ideological over-generalizations which have the effect of blocking class analysis.

In *Solar*, McEwan places his main character—the unsympathetic protagonist that is Nobel physicist Michael Beard, who is habituated to myriad bourgeois compulsions in his personal life—in the boot room and other situations in order to show that not only a physicist like Michael Beard, with his mess of an interpersonal life and promising professional life in shambles, but presumably all of humanity, is caught in "a comedy of failure, of discrepancy between idealism and what happens on the ground."[20] In this sense, Beard can be seen as a caricature of humanism:

while he is a scientific humanist committed to rational understanding of reality, his irrational behaviors (e.g., greed, envy, lust, sloth, gluttony, etc.) undermine his rational intentions to improve the world through science and reason. This focus on a flawed idealism and a compromised set of practices is part of the logic of posthumanism and indeed of an anti-rational metaphysics. Cut off in the imaginary from the material relations in which thought and action are used to accumulate wealth, not meet the needs of all, thought and action become sites of inevitable failure of reason and practice which point to the need to think "outside our biology." Posthumanism sets itself up as a counter to the limits of anthropocentric humanism, and regards all-too-human reason as an inevitable failure, which is seen in the posthumanist logic of McEwan's novel, in which even as he asserts that "our cleverness might win through," his text repeatedly counters reason with unreason, demonstrating the view that irrationality trumps humankind's better intentions. Such a move is more than merely problematic; it is false in the way that ideology is false, in that to dehistoricize the gains and problems of modern society in the age of capitalism, presenting them as instances of a transhistorical human nature, abstracts them from objective historical relations, and ends up repeating the conclusions and consequences of the humanism that posthumanism supposes to supersede. That is, the novel presents as the failings of humanism and human nature what are really the effects of capitalism: both the advancement of the sciences and their domination by capital, and the vulnerability of thinking to ideological formulations. Such abstracting away from the material conditions is metaphysics and metaphysics reproduces ideology.

Michael Beard, as McEwan's figure of this global "comedy of failure," exhibits both great promise (exemplified by his winning of a Nobel Prize in physics for extending the work of Einstein and thus helping to open up new possibilities in photoelectric energy) and also all-too-human compromise (exemplified by his unregulated eating and drinking resulting in obesity and worsening health, his infidelities, his sexism, his unprofessional conduct, and a key instance of plagiarism and crime, which all converge and block the potential benefits of his research). McEwan frames the problem in terms of developing an alternative source of energy which will not change the production and consumption patterns of global capitalism but would potentially allow capitalism to continue for a time as a form of "green" capitalism. This approach makes the crisis of capitalism into a matter of changing over to a new post-carbon energy economy, putting out of view the commodification of labor power (the source of all profit, accumulation, and expansion), and focusing instead on the problem of carbon fuels. Because of how it frames the problem, McEwan's novel argues that this is not only a global crisis (of finding green energy in time) but an all-too-human crisis of compulsions and compromises, and in the end McEwan advances a morality tale about the dangers of unregulated consumption. But the novel makes this a matter of across-the-board consumption, as if everyone consumes likes a Michael Beard, and puts out of view—save for passing references to poverty and how people live without—not only the differences in consumption across status groups, but also and most importantly the class difference in production between the class that consumes labor-power for profit and the class that struggles to reproduce labor-power for its profitable consumption by the ruling class.

Such an approach is symptomatic more generally of the way in which the novel treats problems like consumption and contradictions between thought and action in abstract ways that both generalize and de-materialize human practices and behavior. For the most part, when the novel imagines humanity, it is a largely undivided humanity of civilization. By situating Michael Beard's subjectivity as indicative of a general human nature, the novel universalizes the position of a class fraction of workers (i.e., knowledge workers in the global North) and makes their higher level of consumption appear as a product of human nature. This move completely erases the differences in the conditions of life of people around the world and the relations of exploitation producing these differences. In the middle of the novel, after his actions set in motion both the solar project that may revitalize his stalled professional life and the counter-forces of property and law that will negate this effort, Beard looks down on London from a descending airplane and muses on how seeing the city from above, with ever less visible evidence of nature, gives rise to mixed feelings about the growth of population, technology, and "the blind forces of desires and needs."[21] From this aerial perspective, Beard poetically associates the spread of cities with the mindless growth of mold and fungi. Tracing Michael Beard's metaphysical musings from the view of the city, through the analogy of the mold and fungi, to his imagining of his apartment below, McEwan embellishes the trope through lingering cinematic attention to the growth of mold over dirty dishes in the sink of Beard's apartment down below:

> Over these, over the plates in the dining room, the pans in the kitchen, on the garbage in the pail and spread across the chopping board, and even on the coffee grounds in the dried-out filter paper, there would be vigorous, differently hued fungal growths in creamy whites and soft grayish greens, a blossoming on the abandoned cheese, the carrots, the hardened gravy. Airborne spores, a parallel civilization, invisible and mute, successful living entities. Yes, they would have long settled to their specialized feasts, and when the fuel ran out, they would dry to a smear of charcoal dust.[22]

These musings are metaphysical in that the matter flowing through the arcs of growth and decay forms aestheticized figures embodying a vital logic of seemingly irresistible, objective, autonomous flows which are beyond reason and useless to resist. The comparison of human society to the "parallel civilization" of fungi and mold running mindlessly through the "feasts" and "fuel" to which they are accustomed—and to which McEwan gives lingering attention, to show the nature of Beard's associative musings but also to provide a counter to and effective parody of Beard's more serious intentions—troubles Beard. And yet he "believed in his better moments that he had the answer to the problem. At last he had a mission, it was consuming him, and he was running out of time."[23] Here is the sense of purpose and promise he feels in potentially solving a real problem, or the problem which the novel identifies as simultaneously real and improbable: the problem of climate change as a problem of energy consumption, which a change in energy source could resolve if only human civilization were any better at changing its habits than the parallel civilizations of nature. This is a one-sided, undialectical defining of the problem, locating it in consumption, as if consumption can be isolated from production and the profit priority of capitalism. Consumption, grasped dialectically, exists only in relation to production; production is not only a condition of consumption, but the forms and nature of consumption are determined by the relations of production and the historical development of the social relations and productive forces generally. In capitalist society, wherein production is organized to produce private profit for the property-owning class, consumption is consequently conditioned and determined by profit. Thus the consumption of energy is not the root matter; rather at the base of the "energy crisis" and climate change are the class relations and their impact on the historical development of coal and other fossil fuel industries and the subsequent decisions to prioritize coal, gas, and oil over renewable energy. Capital will turn to renewables when profit is threatened. But whether based in carbon fuels or renewable energy, capital's use of "natural" energy is above all driven by its exploitation of human energy: the accumulation of surplus value in the working day. If, therefore, capital does shift to renewables, they will be used instrumentally: as a means to exploit human labor, leading to ever-deepening social and environmental crises. In the meantime, consumers' use of renewable energy does little to address the depth of capitalist society's reliance on carbon fuels; instead it promotes a new green lifestyle market that continues to depend on a production system based in carbon fuel.

In other words, consumption needs to be understood as part of the social totality of capital. The novel, however, treats the historical conditions of production and consumption as essentially natural phenomena. It does so by making what is a problem that has developed as part of the regime of capitalism into a problem of civilization, and then equating the unplanned and mindless spread of human civilization to the naturally determined action of mold and fungi. The novel, like posthumanism more generally, makes effects of historical and social relations appear as ontological and existential conditions of human life as such.

Indeed, late in the narrative, at the height of the narrative's crises, Beard refocuses on the mission and purpose which, while greater than his individual problems, is being undermined by them, by considering an offer to bail on his doomed solar project and work on "carbon-free nuclear energy":

> The CO_2 levels went on rising and time was running out. There was really only one well-tested means of producing electricity on a scale to meet the needs of a growing world population, and do it soon, without adding to the problem. Many respected environmentalists had come round to this view, that nuclear was the only way out, the lesser of two evils. James Lovelock, Stewart Brand, Tim Flannery, Jared Diamond, Paul Ehrlich. Scientists and good men all. In the new scale of things, was the occasional accident, the local radiation leak, the worst outcome possible? Even without an accident, coal was daily creating a disaster, and the effects were global. Was not the twenty-eight-kilometer exclusion zone around Chernobyl now the biologically richest and most diverse region of Central Europe, with mutation rates in all species of flora and fauna barely above the norm, if at all? Besides, wasn't radiation just another name for sunlight?[24]

The novel dramatizes the view that what is necessary is a new ("nuclear") economy to solve the environmental crisis, making not the social relations of production, but population and (carbon-based) relations of consumption the problem. Further, nuclear power is doubly rationalized as a potential solution in that, unlike coal, it is clean and even in accidents is an opportunity for nature's re-creation in increased biodiversity and mutations. The logical consequence of this line of argument is that there is no need for a collective planned economy of production for need. The shift to nuclear energy, the novel suggests, will solve rising CO_2 levels without the need to address the class relations that systematically put private interests over social need (including the need to preserve the environment). Moreover, it is suggested that this "realization" that the fundamental problems of humanity are linked to consumption is made possible by a "break" with rational ("humanist") thinking. For instance, in a later tense moment, as Beard is engaged in a self-distraction of sexual fantasy, he arrives at a defense of illogical thinking as "a weakness he had to live with," both "infantile" and distracting:

> But similar daydreams—manic moments, brief neural bursts, compacted but cloudy episodes that braided the actual with the unreal and threaded gaudy beads of the impossible, the outrageous, and the contradictory along thought-lines of indeterminate logic—had long ago brought him to formulate his Conflation [the theory that earned him a Nobel]. The poetic, the scientific, the erotic—why should the imagination care what master it served?[25]

Here McEwan allows Beard a defense of anti-rational mental states as the basis of the creativity that is needed in just such a situation as this. At stake here, first, is that the poetic is equated with the irrational and the irrational with the space beyond social and rational determination. Second, the conclusion at which Beard arrives is that being able to solve the problems producing environmental crisis depends on a new appreciation of irrational thought. Leaving aside for now that the turn to the irrational and the anti-scientific are fueling the right-wing attacks on the science of climate change, what is repeated here is the familiar posthumanist claim that to think beyond reason is to think beyond humanism and its anthropomorphism, and represents a "new" thinking outside the social more generally. But, of course, the conclusion that the solution to contemporary capitalism is a new mode of consumption (nuclear energy) is hardly beyond the scope of rational social determination. It "makes sense" from within the framework of consumption, cut off from production. And, as even his own character points out, it reflects a basic argument that has been made for years (more recently in the arguments of Jared Diamond and James Lovelock). Far from being any kind of "marginal" position, it is in fact an increasingly popular argument precisely because it solves the problem of the environment within ideology and at the same time supports the "green" sectors of the capitalist economy. Perhaps just as significant, the novel suggests that the very same notion of "excess" is both the cause and the solution of the environmental crisis: Beard's excessive greediness and irresponsibility makes him emblematic of humanity's inevitable failure; yet it is the excessive nature of the poetic outside of reason that offers the solution to the problems created by greed.

Chakrabarty, in his essay on the anti-dialectical thinking needed to deal with life in the Anthropocene, in fact refers to "anti-rational," or "poetic" thought as a species thinking that will supersede the limits of understanding and critique. It is for him "a figure of the universal that escapes our capacity to experience the world," a universal without "particularities," or what Chakrabarty calls "negative universal history."[26] In this view, to be able to apprehend the apparently unknown and unknowable particularities and causal relations of the Anthropocene is beyond us; we can know only that reason, critique, and even basic understanding will fail. This is apprehension without comprehension. However, grasped dialectically, the turn of reason into unreason (i.e., using carbon-based fuels until doing so becomes an ecological and social crisis) is not due to the nature of reason itself, as in Chakrabarty's metaphysical logic, but is a result of the development of reason under capitalism, whose main objective is accumulation. That is, again, reason does not develop outside of social relations, but is a historical development of the ability to reason, and as such reason is a class practice, distorted into ideological formulations and shaped by economic necessity into counter-ideological and critical modes of thinking and expression. To embrace reason's transhistorical other (the irrational) simply reverses the terms of the binary within capitalism. Scientific reason is needed to understand the relation of carbon to the capitalist mode of production, but it is only through transforming, not just consumption, but the capitalist relations of production, that it is possible to use reason in a reasonable way that, through prioritizing human need, also prioritizes preserving the environment. Chakrabarty and anthropocene theorists however argue that such arguments, which address the relation of human history to climate change, is not "deep" enough. But what they are calling "deep" is the bypassing of human history that actually re-naturalizes environmental crisis. As Malm and Hornborg argue in their critique of anthropocentric theory, "we should dare to probe the depths of social history; not relapse into the false certitude of another natural inevitability."[27] That is, in order to grasp the causality of real problems like climate change as an effect of class society, what people need are concepts like "sociogenic" as the dialectical correction of the metaphysical abstraction of "anthropogenic," as well as science and the counter-ideological practice of critique which produce reliable knowledge of nature and society. The endpoints of *Solar*'s narrative—shattering self-delusion, the loss of potential, failure on all fronts—dramatizes the way of thinking that the novel attributes to a human nature that cannot be controlled and thus should be embraced and celebrated. This is part of what makes Beard the character unsympathetic, but it is also part of a posthumanist logic in which McEwan's fictional discourse trafficks. At the core of the posthumanist embrace of the irrational is the assumption that humanity's failures are rooted in the very nature of humanity and reason as such, which entirely obscures the way class shapes both.

The posthumanist vision of the human as inevitable failure appears in other moments of the novel, such as when Beard, daydreaming while his lover speaks words he does not hear, justifies his failings as "average":

> If he was sometimes greedy, selfish, calculating, mendacious, when to be otherwise would embarrass him, then so was everyone else. Human imperfection was a large subject. Consider just a few of the defects. S-shaped backs that easily buckled, breathing and swallowing recklessly sharing a passage, the infectious proximity of sex and excretion, childbirth pure agony, testicles unwieldy and vulnerable, weak eyesight a general affliction, an immune system that could devour its owner. And that was just the body.[28]

Here behaviors and practices, which are always mediated by social relations, are conflated with natural corporeal weaknesses. Human frailty—weakness and vulnerability—are therefore unavoidable in this logic. The class relations that determine who can afford glasses or surgery to improve eyesight and that prevent treatable diseases and minimize the pain and mortality associated with childbirth for some, all disappear into a naturalized narrative of "human imperfection." This posthumanist representation of the contradictions humans face in class society depends on a dehistoricized and class-free understanding of consumption that in fact mirrors bourgeois class interests. In other words, what the novel represents as a problem of humanism is itself produced by capitalism. Human mortality and corporeality are part of social being, which is a historical condition determined not by transhistorical nature but by historical developments. McEwan's focus on the body, and this generalized failure of the human, shows how posthumanism repeats the metaphysics of humanism. That is, the novel displays what happens when thinking is abstracted from the material conditions: the result is a universalized, naturalized notion of human nature from which class relations have been emptied. To recast the historical relations of class as matters of bodily existence puts an ideological vision in the place of a critical representation of objective reality.

The contradiction within McEwan's novel is that what is first represented as the collapse of the anthropocentric binary of "Nature" and "Society" that justifies doing away with the explanatory dialectical concepts of capitalism and class ends up leading to the very practices within capitalism that are supposed to aid the capitalist class in managing the crisis: new modes of consumption. This is a contradiction in which the appeal to thought and subjectivity as both cause and solution of the crisis of climate and environment puts a false limit on social action by containing it within a closed circuit of consumption. Such thinking is ideological because the thinking that claims thought causes material contradictions is the inversion of reality, when in fact such a claim is a product of the material contradictions in which capital's ownership of the means of production leads to commodity fetishism—as if the latest product, or the greenest form of energy will resolve the problems. The contradictions of the novel—how it cannot resolve the problem it raises, how its irony undercuts its potential for critique—are themselves determined not by thinking but by material practices rooted in the capitalist system. Ideology, metaphysics, idealism—these are material practices in the service of the ruling class. Dialectics, critique, materialism—these are material practices in the interests of the working class and thus in the general interest. What is at stake in re-reading the texts of posthumanism is the historical clarification of the objective class relations in order to make sense of the opposing class imperatives either to adapt to these relations as if they are unchangeable existential fact or to seek to transform them into new social relations free from need.

The anti-dialectics of posthumanism puts class contradiction under erasure in order to assert an ethics of adaptation to what are taken to be universal forces of nature. As such, the class metaphysics of posthumanism is not new, but repeats the humanism it claims to supersede and shares a logic with late twentieth and early twenty-first century critical theories of the social and their differing takes on materialism and the dialectical concepts of totality, contradiction, and change. By identifying the root cause of the global climate crisis in a dehistoricized "human nature" and naturalized subjectivity, McEwan and other writers of the metaphysical narratives of posthumanism enact an ideological logic which leads them to embrace superstructural changes to capitalism that leave intact the contradiction between labor and capital, such that totality exceeds critical comprehension and transformation is deferred to an inscrutable future. Capitalist ideology shifts to a posthumanist metaphysics today both to register the real, material crises caused by capitalism while representing these crises in transhistorical terms that implicitly defend capitalism from critique. This is the effectivity of posthumanism as class metaphysics: producing and circulating new forms of ruling class logic.

This posthumanist logic finds some of its most effective theoretical elaborations in the writings of such social critics as Antonio Negri and Fredric Jameson. Of particular import for grasping the significance of posthumanism as a social theory is clarifying how and why these theorists either reject dialectics as theory of contradiction (i.e., Negri) or repurpose a dialectical vocabulary for coping with global capitalism (i.e., Jameson). While Negri ends up embracing an affirmative ontology that focuses primarily on the resistant creativity of labor (rather than McEwan's concern with humanity's destructive nature), and Jameson reduces all sites of the social to equally "balanced" sites of opposition that are beyond any transformative change, Negri, McEwan, and Jameson all read the social in a metaphysical manner that normalizes (often by ontologizing) the existing class relations of capital. I turn first to Negri's reading of dialectics, to show how such theories provide the means for referring to crises and contradictions in such a way as to protect capitalism from critique.

Whatever the explanatory value of dialectics in the past, according to Negri, the dynamics of capitalism have changed such that its very constituent relations are altered and thus require new theoretical and imaginative means for comprehending these relations and negotiating new terms of existence. In *Empire*, Negri and Hardt write,

> Having achieved the global level, capitalist development is faced directly with the multitude, without mediation. Hence the dialectic, or really the science of the limit and its organization, evaporates. Class struggle, pushing the nation-state toward its abolition and thus going beyond the barriers posed by it, proposes the constitution of Empire as the site of analysis and conflict. Without that barrier, then, the situation of struggle is completely open. Capital and labor are opposed in a directly antagonistic form. This is the fundamental condition of every political theory of communism.[29]

Dialectics as explanation of the existing reality and its relations no longer works because the dialectic of limit and organization changes fundamentally once capitalism is global. Negri and Hardt here argue that once capitalism has expanded to the global level, no longer is capital confronted with "the limits and organization" represented by nation-states, which mediate the encounter between national labor bodies and transnational capital; rather, capital and labor now meet in a global theatre "without mediation" and thus in "direct" antagonism of "multitude" and capital. Multitude is Negri's figure for humanity which must labor to live but for whom each new encounter is an opportunity for spontaneous cooperation. Each singular node of the multitude has within its power the possibility of choosing to cooperate or refuse, and it is this mass individuality—and not class-consciousness of the class-for-itself—which Negri and Hardt posit as "the fundamental condition" for the transformation of capitalism into communism. This is a theory of the objective relations of class as secondary to the subjective choice which the members of multitude have in each new encounter. As I will argue, this is a theory which claims to address objective reality from a radical standpoint, but, by making subjectivity primary, it actually denies the objective determinations of class and exploitation, and as a consequence is ultimately not a theory for transformation but a theory for adaptation to worsening exploitation. In fact, its reading of subjectivity as ontology articulates in a different idiom the posthumanist metaphysics by which history is submerged in a post-class affirmation of "creative energies" and the "passions of life."

In his rejection of dialectics as a means for grasping material reality and social change, Negri puts forward an ontological theory of the multitude, singularity, the void, the monster, and other figures. According to Negri, while Marxist theory is based on binary codes which "waver between (capitalist) subject and (proletarian) object of exploitation," Negri's theory breaks free of such formal restrictions in that his terms carry an "ontological indexicality [. . .] much deeper" than Marx's and thus do not waver between the poles of conventional binaries but rather pass through to an active space "*between subject and subject*."[30] Negri writes that Marx's work "beyond the dialectic he has inherited from Hegel" is limited to "a logical-political point of view" and is thus implicated within the traditional logics of capitalism and western metaphysics, whereas what is today "necessary instead [is] to dig deeper and foreground ontology," which goes beyond traditional conventions of thought and enables the force of refusal and resistance that forms a forceful alternative outside the "singular communication between capital and living labor."[31] It is significant that Negri claims that his ontological logic goes "deeper" in its account of reality and change than does dialectics, and thus offers a more "truthful" depiction of reality. In going both beyond what he presents as the limited binary of capital and labor and thus beyond class as subject and object into a third space between "subject and subject," Negri bypasses the social as a space of relations understood dialectically as contradictory and emphasizes these relations as expressive of a basic ontology that precedes class as a historical relation. Negri argues that taking up an ontological viewpoint enables movement beyond the limits of dialectics and its static "wavering," and that moving beyond the inherited conventions is required to address the complexity of human relations today. This is because, according to Negri, beyond Marx's theory which was explanatory of an earlier moment in modernity, capitalism today is singularly an expression of the multitude, the autonomy of labor from capital gathered together as "free mass intellectuality"[32] whose immanent self-movement produces the real.

Following out a logic parallel to that taken up by Jameson and other theorists of globalization in which the increasing "computerization" of industry, finance, communications, media, and many aspects of daily life is supposed to bring about a new set of ontological relations in the sociocultural field (what Negri and Hardt call "postmodernization"), Negri argues that the binary of class, which Marx regards as an objective relation between the capitalist subject/agent and the laboring object, has shifted through "postmodernization" to become an encounter between subjects. Negri essentially rewrites the contradictory binary of class, in which capital determines the terms of the wage-labor exchange, into a "complementary dyad"[33] of subjects, free to cooperate or refuse, and apparently without exploitation as a condition for producing value. Negri and Hardt argue that labor takes on new forms today, such that labor in "informationalized" industrial manufacturing "mixes with and tends toward immaterial labor," and that "analytical and symbolic" labor (such as in education, finance, and engineering) and "affective labor" (such as healthcare and customer service) are forms of immaterial labor which are increasingly the dominant form of labor.[34] They write that "Immaterial labor immediately involves social interaction and cooperation" and that "the cooperative aspect of immaterial labor is not imposed or organized from the outside, as it was in previous forms of labor, but rather, cooperation is completely immanent to the laboring activity itself."[35] Thus, rather than regarding labor as "variable capital," which for Marx represents the exploitative relation between capital as the determinant over the expenditure of labor-power in the wage-labor relation and labor as the possessor of labor-power which can only be expended on capital's demand, Negri and Hardt regard labor's cooperation as voluntary and spontaneous, subject to the will of the worker. This notion that "the cooperative powers of labor power (particularly immaterial labor-power) afford labor the possibility of valorizing itself" as an autonomous subject and no longer the object of capital is an attempt to rewrite the binary of class and the fact of exploitation into a plurality of singular subjects with freedom to choose to cooperate or not. Put differently, Negri and Hardt do not so much get rid of the binary of class and the fact of exploitation as they redescribe it to emphasize one half of the social contradiction. Their claim that "In the expression of its own creative energies, immaterial labor thus seems to provide the potential for a kind of spontaneous and elementary communism" alludes to the Marxist idea that the development of the productive forces is not only a condition for the productive exploitation of labor by capital but is importantly a condition for the working-class's developed consciousness of itself as a potential class-for-itself.

Certainly, the potential for social transformation to new social relations exists in the development of the productive forces; however, what Negri and Hardt leave aside is that it is not only in the development of the productive forces that society develops but also in the relations of production, which are historical relations that enable production until they become a disenabling force (what Marx figures as "fetters"[36]) and thus of necessity are revolutionized into new relations. But Negri and Hardt disappear the exploitative relations of production under capitalism when they rewrite the class relation as not one of objective exploitation for the production of value but one of subjective interaction and potential cooperation with value produced out of labor itself as an "expression of creative energies." Negri and Hardt, that is, emphasize labor as a source of value, but make the claim that labor is now free to produce and claim its own value, outside its relation with capital. But this is neither communism nor an accurate portrayal of the reality of labor today. For example, software engineers who cooperate in the creation of new operating systems are not free to claim their designs as their own value; in fact, the terms of their cooperation and intellectual labor are rigorously determined and conditioned by legal restrictions through which the capitalist enterprise claims ownership. Negri treats the objective relations, which in reality are the objective property relations which enable exploitation, as networks of subjectivity. The problem, from the standpoint of class and transformation, is that Negri turns the objective conditions of social being (historical ontology) into subjective conditions of free choice (transhistorical ontology), rendering the exploitative relations of class into subjective self/other relations of communication, interaction, cooperation, and choice that only has to be seen as free to become free. Negri's reading of global capitalism as "new" erases classes as an objective relation and turns communism into a state of being (ontology) and not the movement to bring about the end of class. Transformation, consequently, becomes a guide for self-valorization within capitalism rather than the abolition of exploitation. Negri thus articulates in a different idiom what Chakrabarty and other posthumanists term "deep history," which is an anti-dialectical mode of historiography. Deep history, according to its proponents, is history that goes far beyond the limits of human history. Like Negri's prioritizing of ontology, deep history folds social relations into nature, suggesting that theories of the social and class are not "deep" enough:

ontology is "deeper" than capitalism for Negri, and ecology "deeper" than class relations for Chakrabarty.[37] But against dialectics and the logical analysis and reconstruction of cause-and-effect in the natural and social systems, posthumanist texts and theories emphasize an anti-dialectical, and indeed more generally anti-rational way of being that is not "limited" to or by thinking. This is ultimately an adaptation to the effects of the problem and not at all a solution that goes to the cause.

Negri argues that dialectics favors the objectivizing of labor by the subjects of capital, whereas ontologics favors the encounter of different subjects and views change as a matter of "refusal" and "resistance" in the pursuit of self-valorization. The advantage Negri claims for ontology over dialectics is that the "communication" between "capital and living labor" is not limited to a "singular communication" between the commanding subject (capital as agent) and the structurally delimited object (living labor), so that labor, inhabited by the ur-force of "the monster," can "separate from capital"[38] and choose to refuse and resist. This approach de-emphasizes the structural dynamics of capitalism and puts much greater emphasis on the notion of multitude as a (class) agency of self-affirmation. In *Empire* and *Multitude*, Negri and Hardt suggest that to emphasize objective over subjective force is to see society as static, and they call for a change in our way of thinking, seeing, and acting. Negri develops this argument in *In Praise of the Common*, where he states that in the "objectivistic" version of Marxism, "there are always two opposing forces—capital and labor—that can never reach a synthesis, at most, they can temporarily reach more or less stable equilibriums."[39] What Marx represents as the relation of "the material world" and "the ideal" (i.e., "the ideal is nothing but the material world reflected in the mind of man, and translated into forms of thought,"[40]) is for Negri a fundamental contradiction, in that he translates the reality of capitalism into a formal problem of Marxism, basically asserting that dialectical materialism is flawed because it depends on binary contradictions, imposing the form of contradiction on reality. In rejecting class as an objective binary of the dominant subject of capital for which labor is a necessary object for the creation of value, Negri does not supersede the binary logic of the class contradiction but instead rewrites it in the non-contradictory terms of complementarity and cooperation. The result, Negri claims, is that "the dualism constituted by these two figures [of labor and capital] is not at all a metaphysical dualism; it is, rather, an empirical as well as a materialist dualism, which cannot be sublated in any way."[41] In so doing, Negri maintains a binary, but evacuates it of its social contradiction, turning it into a figure of transhistorical ontology. In other words, he makes the "dualism" of labor and capital *static*—the very condition he claimed to avoid by abandoning the Marxist dialectic, replacing objectivity with subjectivity.

This eternalizing of the capital/labor dualism is the effect of his ideological theorization of subjectivity as ontology, even as he seeks to give his theory a historical and material gloss. For instance, while by insisting that his re-theorized capital/labor dualism is "not" a "metaphysical dualism," Negri suggests that capital/labor is not just an idea or figure but is, as he says, "empirical" (knowable) and material, or in other words, real, he is quick to emphasize that this real dualism—which Marx theorizes as the central contradictory dynamic of capitalism as a historical totality—*cannot be overcome and transformed into some other different relation*. Indeed, as Negri puts it, the real dualism of capital/labor "is determined over and over again—and each and every time it is redetermined by exploitation. And exploitation is always a machine for the production of subjectivity, is always a subject formation."[42] In this rewriting of the ontology of class as not objective but subjective, Negri makes exploitation significant not for its structural, material role as the source of both private profit and impoverishment through the wage (the same contradiction expressed at a different level), but rather for its subjectivizing function in recreating ("producing") perpetual opportunities for authentic being. Negri's emphasis is that it is in subjectivity that the multitude exists as the resistance to objective relations and thus gets outside capital/labor dualism and its false limit of "dialectic." The outside—or the potential for change—is not in changing the objective relation but in affirming a different subjectivity. While on the face of it, Negri's embrace of the notion of workers' autonomous subjectivity may seem empowering, this move locates the resolution of class conflict in reactive subjectivity and marginalizes the objective class contradiction as a feature of the subjective relations. Far from avoiding metaphysics, Negri's theory of subjectivity is metaphysical through and through. It erases class as the historical and objective structure shaping subjectivity under capitalism, and embraces the notion that subjective acts that appear separated from structural contradictions are ("ontologically") "free." Negri's theory withdraws into a resisting subjectivity that will appear to be "outside" class to the extent that it can assert and affirm that the subject exists as resistant being outside the exploitation that perpetually redetermines "the real dualism of capital/labor." The effect is to produce a "resistance" which changes nothing and aids in the continuation of capitalist relations.

As I have marked above, Negri ideologically dissolves the material contradiction between capital and labor and, therefore, erases exploitation. He then proposes that we live a time of "static equilibrium" of capital and labor;[43] that is, a capitalism without antagonism, contradiction, or exploitation. This formulation represents the class relation at the base of capitalism as static in that it cannot be overcome but is an existential space in which singularities meet and choose whether and how to cooperate. Thus in "a world with no mediations,"[44] Negri regards the "complementary dyad"[45] of singular and common as more explanatory than the so-called, outdated Marxist theory of class. Singular and common, unlike Hegel's self/other or master/slave, or Marx's binary of bourgeoisie/proletariat, is, according to this narrative, not to be seen as a matter of definitive negation, but rather as one of confronting the void.[46] For Negri, there is "no sublation" in the complementary dyad of singular/common in that "complementarity is not a dialectical relation," and complementarity compels a "transformative rather than a dialectical synthesis,"[47] that is, one of reconciliation rather than contradiction. By this logic, which is premised on narrow limits of possible alterations within capitalism rather than transformation of the relations of capitalism, revolution is no longer negation through the sublation of the current class contradictions but rather is to be found in the "love"[48] of "being-together,"[49] an affective and ethical, rather than an objective and material, shift toward a classic liberal celebration of difference without opposition. What Negri seeks is not sublation (transformation) but rather a "metamorphosis"[50] that is ultimately an ethical change in subjective relations, not negation of the present objective relations. Negri rejects materialist dialectics for the very reasons that it is needed today. In his ethical approach, he rejects the possibility of transforming the basic class contradiction of exploitation, instead converting class relations which emerge historically into a universal state of being defined as singularity and multitude against the Void. By this logic, the singular is the radical "other" in itself that is like no other and opposed to none, and the common is the multitude of singularities (as Negri puts it), or the being in common with others who are not opposed.

Negri finds great power—agential power of self-affirmation in refusal and resistance—in this re-subjectivizing of the working class and all those who are objectivized by capital, to the point that he claims that "in the capitalist regime, when monstrous resistances have grown to such a degree of physical consistency, any dialectic and any linear transformation [. . .] has become ridiculous and impossible."[51] In this representation of contemporary capitalism, the working class has become extraordinarily powerful (a claim which not only inverts the relation of capital and labor at a time or unprecedented gap between exploiters and exploited, but also, implicitly, suggests that if the working class are not free from the control by capital, it is their own fault). At the same time, this power of the working class cannot be understood dialectically as leading to the negation of the capital/labor relation. Dialectic and the notion of total change (transformation, sublation, *Aufhebung*) that has a clear line (linearity) of cause-and-effect is today "ridiculous and impossible" in that the relation of capital and (living) labor is no longer—in the era of immaterial labor—a relation of mediation but rather one of resistance.[52] Indeed, as Negri puts it, "the relation between capital and proletariat (the *multitude* of those who produce and are exploited) is founded upon an ontological imbalance in favor of the proletariat."[53] Here is a point where Negri emphasizes a potential power of the exploited class, and yet he immediately here and elsewhere denies class as a socially contradictory structural position (objective identity) and instead proposes multitude as an ontological figure "deeper" than any concept of class, wherein "resistance is no longer only a form of struggle but a figure of existence."[54] That is, multitude displaces class as force within history, such that "class struggle" (which for Marx is the dialectic of historical change) is rewritten as "a figure of existence": the ontological "resistance" of the multitude. This is an existential rewriting of class, which displaces the historicity of class as an objective social relation with a posthumanist subjective (experiential, existential) relation. But class, grasped dialectically as a material reality, is a historical relation between the owners of property and those who own only their labor-power; it is primarily an objective relation at the site of production, which is the basis for the mediation of class into various superstructural forms, including ideological ideas of class (such as Negri's ontological metaphysics of class as multitude) and counter-ideological ideas of class (such as revolutionary ideas in struggle and the production of class-consciousness). Negri's concept of the multitude—the free association of various singularities—makes resistance into a way of being, not a historical force for becoming, sublation, transformation.

As part of this metaphysical understanding of ontology, Negri denies that historical cause and effect can be known because of the chaotic and subjective nature of reality and our position within it. He writes, "[Nor], finally, is it possible to grasp the efficient cause, because what is before us is not only a process, an accumulation, but above all a transfiguration, an explosion."[55] This notion that reality is "an explosion" beyond explanation, is part of Negri's rejection of dialectical explanation. Thus, in order to account for reality, history, and change, Negri turns to figures such as "monster," he claims, because such images and tropes "tell us, outside any dialectics, what 'to be monster' means today: a common subject, a collective force, a being other";[56] that is, such figures of thought and language—like the "carnivalesque masses" of human (and other) bodies, with their "multitudes of desires"—"will exist in their resistance to any attempt to manipulate their power besides the power to resist, they reclaim also the plenitude and the richness of the passions of life."[57] For Negri, the subject of labor is no longer subjected to capital; it has its own power separate from the relations of production. This leads not only to the rewriting of history in terms of a mutational logic that defies dialectics and causality more generally, but the advocating of labor's "exodus" from capital.[58] In other words, like Michael Beard's distracted daydreaming and following of desire and drive, such figures apparently "resist" falling into objectivity (which purportedly negates agency) and instead affirm a subjectivity of refusal and escape. This is a theory of history and social totality which on the surface elides class as a factor but nonetheless expresses the contradictory class reality by representing as a generalized "resisting" agency of the common, multitude, singularities, what is in reality bourgeois privilege; that is, the ability to refuse and escape has a material basis, with those removed from the necessity of labor better positioned to do so than those for whom the stakes at the site of exploitation are higher. The objective reality of class is that privileged individuals have greater social and cultural agency than do those individuals who have little if any distance from necessity. Negri, like McEwan's Beard and like Chakrabarty's equating of all people under the abstraction of species, produces a theory of class which abstracts the objective contradiction of class and recasts class as a subjective agency. This is a metaphysical theory of ontology which strips being of its social determinations only to rewrite these determinations as general conditions of all life. By absorbing class into these tropes and notions, Negri separates subjective ontological "power" from the objective historical structural totality of class relations in capitalism. Thus, his figuring and theory result in the production of subjectivity as a removal from the struggle to transform the social relations to a seemingly autonomous space of resistance.

That Negri's theory of ontology and subjective agency is really based on the deep de-historicization and de-materialization of labor and agency is perhaps made even more explicit in his theorization of "the nonontological absolute" of "the Void," against which all creativity, affirmation, and production (of subjectivity) occurs. He writes, "Each of us builds on the void: any form or presence is a bridge that we cast over the void."[59] The void is present in our every action in that "eternity is chance" and we cannot know ahead of time the outcome of any risk we take, even as we can in fact know that "what we build on the void cannot be undone"[60] and this is history. This, Negri asserts, is why "the dialectic is shit!"[61] That is, for Negri, the dialectic is inadequate at accounting for change because "It always tries to recombine and reconcile everything, to bring everything together, to make everything fit together, to force everything to make sense."[62] By treating dialectics as a forced formalization, Negri rejects the dialectical materialist concepts of totality and Marx's dialectical argument for "unity of the diverse," whereby what ultimately accounts for difference is unity within historical relations, relations which can be studied and known, relating the concrete and the abstract. But Negri here rejects as error the attempt to interrelate the apparently different, in favor of seeing reality as a space of non-related or unmediated singularities with no necessary developmental relation. Thus for Negri, Marx's dialectical theory of totality as "the unity of the diverse"—which can either be grasped *dialectically* through analysis of the basic relations of different facts and thus reconstructed in thought and discourse so that totality is apprehended as "a rich totality of many determinations and relations," or accounted for *metaphysically* in what Marx calls "a chaotic conception of the whole"[63] —is inadequate because its attempt to create a coherent relational explanation is done "so as to please everybody, to make everybody happy, and to pacify us all!"[64] That is, Negri is suspicious of the idea that dialectical explanation can be coherent given that he regards dialectics as a class-bound theory that cannot get outside of capital relations, unlike ontology which (following posthumanist theory's rejection of "human" "constructs") goes beyond history and class relations to nature and the Void.

In the introduction to *Grundrisse*, Marx critiques the method of political economy for explaining totality in ideologically reductive, formalist terms. Clearly any theory and model will be a reduction; the key is that the reduction be based not on preconceived abstractions (ideals, forms) but on a working-up of the basic facts, arrived at by working-down from apparent abstractions to the "simplest determinations"[65] and then reconstruction back up to a coherent, explanatory, reliable, actionable account of the structural totality in motion. Using the example of population, Marx demonstrates that population is an abstraction, "a chaotic conception of the whole,"[66] if not grasped in terms of class and that class is "an empty phrase" if the chain of mediations down to the most basic relation are not in view. Once the analyst has cut through the abstractions, followed down to the "simplest determinations" which can be known as facts or taken as reliable givens, then one follows back up the chain of mediations once more to "population," "this time not as the chaotic conception of the whole, but as a rich totality of many determinations and relations."[67] The goal is not to know population as an abstraction but as a concrete, that is, "concrete because it is the concentration of many determinations, hence unity of the diverse."[68] This dialectical premise—unity of the diverse—is not an assumption but an inference at the core of materialist dialectics: what exists now exists in relation to everything else that exists, by necessity. That is, what exists now has a history of relation with everything else. They are in dialectical relation and these relations can be known dialectically, which means following the chains of cause and effect, of mediation, of complex historical relations.

Whereas Negri explicitly rejects dialectics as an outdated method, Jameson "rehabilitates" dialectics by turning it into a methodology of "on the one hand/on the other hand" thinking. That is, Jameson takes dialectics, which for Marx is an intensification and eventual overturning of social antagonisms and conflicts, and makes it into a formal machine of social stabilization. In *The Political Unconscious,* Jameson calls for a "dialectical criticism" by which to show the mediation of the economic through the cultural. The nature of this mediation and thus the character of his dialectical criticism is articulated in *Postmodernism,* in which Jameson proposes that, owing to developments in technology, media, and the global market, "everything in our social life [. . .] can be said to have become 'cultural' in some original and yet untheorized sense,"[69] thus requiring a new "dialectical" mode of cultural apperception which he terms "cognitive mapping."[70] In these works, as in his more recent *Valences of the Dialectic* and *Representing Capital,* Jameson enacts a formalist reading by which the traces of totality can be revealed in texts in the form of a balance between the tropes of Ideology and Utopia, with "Ideology" referring to the way in which a given text expresses partisan class interests, and thus negates opposing class interests, and "Utopia" referring to how the text affirms a resolution of class conflict in some future peaceful and equitable state. In *Valences,* Jameson frames this textual dynamic as a tension between ideological "replication"[71] of existing class relations and the wish for Utopia.

In order to "dialectically" grasp the complexity of a text in the contemporary historical situation of late capitalism, Jameson advocates an ideological analysis "grounded in the union of opposites, and positing an ambivalence of replication and Utopia."[72] Jameson's version of "dialectics," as sketched here and shown more fully below, preserves much of the form (e.g., vocabulary) but evacuates dialectics of its historical and material content, thus appearing to many as a materialist theory but producing not a theory for transformation but a theory for adaptation to existing class relations. Jameson's reformism can be seen in his equating of contradiction and binary opposition. In *Representing Capital,* Jameson writes that

> the fundamental notion of contradiction itself [. . .] is identical with the unity of opposites, opposites which no longer need to be identified with and labeled as positive and negative, inasmuch as the dialectic means a perpetual changing of places between them and a perpetual transformation of one into the other. To be even more technical about it for a moment, this means that the famous "identity of identity and non-identity" is also the same as the "non-identity of identity and non-identity"; at which point I fear that philosophy really is at an end in its older sense, and in which it also becomes impossible to "decide" whether capitalism is positive or negative, good or evil, inasmuch as, according to the Manifesto, it is both at one and the same time.[73]

Jameson plays with the concept of the unity of opposites, making the unity of good and bad, for instance, into an equation by which shared identity or unity—in this case, moral value—and difference within unity—here, opposition—undo each other philosophically and the distinctions break down. Jameson is reading Marx's assessment of the advances represented by the bourgeoisie as a revolutionary class and the developments in production within capitalism as if Marx simply "balances" the revolutionary aspects of bourgeois development against its counter-revolutionary measures; or in other words, as if Marx suggests that capital's violent exploitation of working people and their ideological containment is "balanced" by capital's technological progress. In fact, Marx treats the dialectical contradictions at the root of capital as signaling the necessity of the overthrow of capital by labor, and the development of a society in which technological advances are shared by all, not used to accumulate private wealth. But Jameson allows that "the dialectic means a perpetual changing of places" between opposites, and "a perpetual transformation of one into the other," suggesting that in making sense of two sides of a contradiction, dialectics involves undoing the conceptual distinctions—between good and bad, or between bourgeoisie and proletariat—which are necessary for understanding capitalism and thus the dialectical thinker is unable to decide on either philosophical or practical matters. By this logic, one is either paralyzed, being without grounds for decision, or one can choose according to "preference," given that because binary oppositions such as good/bad are part of "our common-sense everyday common language," then "we are left with the opposition, which we can emphasize as we prefer."[74] What decides this preference? Jameson claims that

> the choice between a "good" description of capitalism (as constant revolutionizing and innovation) and a bad one (as exploitation and domination) is in fact a political choice and not a logical or scientific one: a choice that must be made in function of the current situation, and whether people can be politically energized by the negative—anger—or the positive—hope.[75]

Jameson's separation of politics and social decision from reason and science recalls Chakrabarty's argument that politics has never been based on reason, and thus affective response—here anger or hope—is the basis of political decision. This repeats the posthumanist logic of McEwan's *Solar* as well, in which affect and so-called natural drives are at odds with rational decision and science. In all these cases, emotion itself is severed from material relations, making, for instance, an individual's "anger" at the effects of exploitation an individual or spontaneous event disconnected from social conditions, an approach to emotion which also prevents analysis of the role of ideology in redirecting attention away from the system that systematically causes anger-producing practices. Along these lines, Jameson's understanding of contradiction, which is present here in the opposition of the revolutionary and exploitative aspects of capitalism, represents not as an objective state of affairs which can be explained and made subject to social decision through science and reason, but as a totality with two main valences about which one is either ambivalent or can be swayed ("politically energized") by negative or positive affects which exceed reason. Thus, contradiction, which for materialism is an instability at the base of change, is rewritten here as a self-contained ambivalence that reflects differences but does not drive change.

Jameson's version of dialectics is based on his claim that "the dialectic came into being to handle a strange and unparalleled historical situation"[76] in which continuity and change are simultaneous: a new mode of production and new social relations emerging out of former relations, notably the transition from feudalism to capitalism. There are social groups, classes "unique to our mode of production," yet they are distinct as classes. From this tautology, Jameson concludes that "The dialectic is then a mode of thinking able to combine the singular and the general in a unique way, or better still, to shift gears from one to the other and back again, to identify them in such a way that they remain different."[77] The dialectic, for Jameson, is a way to think through the commonsense "in a unique way" that is both a temporary "interruption"[78] of commonsense everyday thinking and the unavoidable reification that fixes things rather than apprehends them in flux. On these terms, the dialectic temporarily undoes the commonsense of capitalism which is then recontained by ideological replication. Because an interruption is temporary, by this logic, it must be repeated since language and ideology exist through repetition. Such repetition, Jameson posits, can only be a "singular" and "formalist" intervention in that

> the dialectic is not a universal or generality of that kind, of which there might be examples. Each dialectical moment is unique and ungeneralizable, and this is why we are able to describe what is dialectical only in terms of various shapes (unity of opposites, contradictions, etc.) and not in terms of abstract concepts.[79]

What makes the dialectic "historical," by Jameson's logic, is both that dialectics emerges as a way to make sense of a newly emergent state of social relations but also that the dialectic is specific to a particular historical moment, although just how long a moment is in duration is not explained. Because history is change, by Jameson's logic, the dialectic changes in its content from moment to moment within a discrete totality defined by a specific mode of production, but the "shape" or formal aspects of dialectics persist across moments, presumably within the mode of production of capitalism, or so long as the historical conditions necessitating the dialectic persist. This theory of the historical necessity of dialectics turns the dialectic from an explanatory mode for critically grasping mediations and relations into a formalist and descriptive mapping of particularities and generalities that explains nothing, but provides a vast field of differential relations. By this logic, Jameson considers Marx's *Capital* as "itself a unique historical event,"[80] an instance of dialectics in that the text develops forms (i.e., dialectical "shapes") for making capitalism intelligible in its historical moment. *Capital* is "dialectical," according to Jameson, in that the text uniquely recognizes that "capitalism is not a concept but a historical phenomenon (with its twin faces of structure and event)."[81] Jameson severs concepts as generalized forms from the historical phenomena and their relations, and renders historical particularities into "unique and ungeneralizable" objects which can be described but resist adequate description.

Such a dialectics of ambivalence, like Negri's "static equilibrium," reinforces an ideological limit on both the ability to imagine social change and, equally important, on critically apprehending the relational dynamics of the capitalist totality: that is, these left theories, in their displacement of materialist dialectics, abandon explanations of exploitation as the cause of the current crises of global capitalism and instead settle for complicated descriptions of the surface realities that are to be "recognized" and "appreciated" through more refined and mature ethics. As such, it is not a surprise that Jameson's reading of capitalism, like Negri's, takes as its starting point the presupposition that the base/superstructure distinction no longer applies, and that "capitalism's third age"[82] (i.e., the age of neoliberalism) has made *all* thinking ideological and thus greatly reduces if not eliminates entirely the possibility of a dialectical position from which to enact ideology critique with any but "cynical reason."[83] In other words, Jameson's "dialectical" move here is to attest to the Utopian wish of dialectics as means by which to provide a reliable critical account of the dynamics of global capitalism as a social totality (which he presumably thinks his work attains) while simultaneously putting into question the possibility that capitalist system leaves any area of material subjectivity open for a genuine critique. In other words, Jameson's ambivalent dialectics here is a matter of both marking the necessity of critique and stating that such a critique is no longer possible because there is no longer any basis outside ideology from which to stage a critique. Critique, in effect, becomes a cynical performance.

Jameson's theorization of dialectics conflates totality and ideology, as if the understanding of the social as a totality means there is no outside to ideology. But dialectically and from the standpoint of an unambivalent materialism, any critique of totality can only come from within totality, that is, from the submerged side of the social contradiction, still inside the totality. Indeed, ideology and counter-ideology are mediations of the basic contradiction of class, although for Jameson, ideology has become a generalized trope, no longer an explanatory concept. But the outside to capital exists within the relations of capitalism as capital's opposed other. It is the other side of the basic social contradiction of capitalism as a historical totality: labor. The standpoint of the working class is the outside because, for instance, as Marx argues in the *Communist Manifesto,* "The proletarians cannot become masters of the productive forces of society, except by abolishing their own previous mode of appropriation, and thereby also every other previous mode of appropriation. . . . The proletarian movement is the self-conscious, independent movement of the immense majority, in the interests of the immense majority."[84] If ideology seems increasingly to dominate the field of theory and culture, this is not because the conflict driving capital has disappeared, as some would argue. Rather, class contradiction and conflict have developed in such a way that theory itself has become commodified. If theory is to be materialist dialectics and not formalist metaphysics, then its dialectical task is not to accept the ideological disappearance of the outside, but rather to make the proletariat visible and intelligible, as a dialectical force of change, in order to develop the self-consciousness of the working class as a class with a "mission to destroy all previous securities, and insurances of, individual property."[85] But Jameson's "rescue" of the dialectic renders dialectics into metaphysics.

Jameson's celebrated reading of Wal-Mart demonstrates some of the concrete consequences of his treatment of the dialectic. In his reading of Wal-Mart as a social phenomenon, Jameson characterizes the dialectic as "an injunction to think the negative and the positive at one and the same time, in the unity of a single thought, there where moralizing wants to have the luxury of condemning this evil without particularly imagining anything else in its place."[86] This is a formalist "injunction" to hold together "in the unity of a single thought," as with Keats' negative capability, both sides of a social fact like Wal-Mart. But such thinking, like all formalism, displaces dialectical unity, arresting the moving contradiction, removing it from history, and making it into a "static" binary cut off from its origins and outcomes. Jameson is suggesting that to understand the social phenomenon of Wal-Mart is not only look at both sides, but to have the "luxury" of "moralizing" (not the necessity of critical analysis) without being compelled by dialectical necessity to imagine an alternative. The ambivalence here—an expression of the class contradiction that Jameson evokes and then suspends—is in the possibility of "condemning" Wal-Mart but not being compelled to make predictions about the future. Jameson is suggesting that attempts to name and represent the future—the imagined alternative to condemnable Wal-Mart—are always going to fail because we can only accurately imagine the past and the present and will in fact fall into projecting what we know onto the future. Only the subjects of the future can (try to) represent the future. Jameson's ambivalence about wanting to value present judgment and also wishing for better judgment suspends critique as a social force, leaving it to shuttle between appreciative recognition of the affirmation of capital and critical exposure of the negation of labor. Social phenomena, then, appear as expressions of both sides of the current class relations, such that, as Marcial Gonzalez claims, Jameson's "'arrested dialectic' represents class struggle at a standstill."[87]

Wal-Mart, grasped "dialectically" by Jameson, must be seen for what it is, in all ambivalence: able both to "hold down or even lower prices and make life affordable for the poorest Americans" and to be "the very source of their poverty and the prime mover in the dissolution of American industrial productivity and the irrevocable destruction of the American small town."[88] To dialectically appreciate Wal-Mart, by this logic, is to apprehend Wal-Mart for its different valences—marvelous and destructive—that is, to trace out the replication of ideology and the hope for Utopia in the present configuration, rather than negate its negation of the agency of labor, in the overturning of capitalism. Dialectics thus is flattened into what Carolyn Lesjak, after Jameson, celebrates as "recognition"[89] or "learning what we already know."[90] Dialectics, in Jameson's logic, is no longer a transformative mode but an appreciative one. Jameson's is a debilitating class theory that both reveals the potential for transformation and yet cynically defers such transformation as always potential but never kinetic or realizable.

On the contrary, the point of the dialectic is that for Marxism, contradiction is the motor of historical change in that one of the two opposing forces in a contradiction dominates the other. For example, in the contradictory relation of capital and labor, while labor is the source of wealth, capital appears as the main source of value. The dialectic is the historical correction of this false appearance and accounts for why such appearances are the effect of the relations of production in which the stripping of labor of all property except labor power is the source of this appearance and, at the same time, establishes the basis of the negation of the negation that will overturn exploitative property relations once and for all. However, Jameson and Negri, like others on the left, accept the false appearance by suspending the conflict between labor and capital within the social contradiction of class exploitation as if it were an unresolvable contradiction. Despite their formal differences, both displace contradiction with balanced dyads or static binaries, respectively, such that the class contradiction appears as a perpetual antagonism. In this sense, Jameson's ambivalent dialectics achieve much the same effect as the antidialectics of Negri, which is the replication of the ideological notion that class and exploitation, and thus all the suffering and inequity that follow out from these become, on Negri and Jameson's terms, unchangeable realities to which we must continually readjust our expectations. Thus, Negri and Jameson, as leading theorists of the left, provide the theoretical basis for posthumanism as class metaphysics.

Like Negri's displacement of dialectics through the substitution of reconciliation for sublation, Jameson's suspension and deferral of sublation, and thus of transformation as the outcome of contradiction, allows the recognition of oppositions as differences while positing them as either unchangeable or as unnecessary to change. Whereas sublation, materially grasped means the supersession of a contradictory situation into a fundamentally transformed state, what Negri proposes instead is a bringing into balance of singular and common through "love," and is thus little more than a reconciliation of classes at the level of ideas. Such theoretical moves are akin to the posthumanist ontological logic of changing consumption without changing production relations and indeed without accounting for the determination of consumption (status, inequality) by priorities of production.

Contra such theories which reproduce a self-perpetuating social defeatism in the face of necessary social change at every level of the contemporary social totality of global capitalism, from production priorities and the class relations of exploitation at the base, to how these basic relations and the decisions made to preserve them are mediated all through the superstructural forms, Marxist theorists working with the materialist dialectic build on Marx's insight that "It is not the consciousness of men that determines their being, but, on the contrary, their social being that determines their consciousness."[91] The divided and contradictory forms of consciousness of social being (e.g., the class antagonism of the ideology of metaphysics to historical materialist dialectics) do not originate in consciousness, but are mediations of the basic class contradiction which defines social being in capitalism: the exploitative production relations, and the inevitable structural conflicts which exploitation engenders, are mediated all through the social totality. As John Rees argues, in his explanation of how Marx changes Hegel's dialectics into historical materialist dialectics, "Mediation is no longer a peaceful process of reconciliation but the elaboration of the different forms in which the central contradiction of the age is played out in every aspect of social development."[92] "No longer," in that, materialist dialectics, against idealism and metaphysics, is explanation of concrete reality and not merely the working-out of forms and abstractions. Given the historical urgency of the unmet needs of the majority of the world's people, and now the development of the social contradiction of capitalism into the world-historical and global crisis of climate change and all its diverse mediations, theorists of society and its relation to nature must "no longer" frame the historical contradictions and crises as unchangeable realities to which "humanity" must adapt, but must subject such metaphysics to rigorous critique and dialectical explanation.

The influential theories of Negri and Jameson, among other theorists, form the basic logic of posthumanism, its metaphysics and class logic, thus serving the ruling class in the short term, by distributing the managerial ideological discourses for accepting the what-is of capitalism not as a dialectical becoming and passing away, but as an existential condition of life to which people must adapt. At a time of burgeoning global crises—social crises of poverty, disease and early death, alienation, war, and other forms of class violence—as well as the sociogenic crisis of climate change, overall and in its regional iterations, what is necessary is a clear assessment of the causal relations. For it will be in the explanation of cause and effect that both the problem and its potential solutions will be best understood and made subject to social action. Posthumanist metaphysics and the counter-dialectical theories which undergird them are managerial theories enabling the ruling class to continue its domination of social planning and decision. This is demonstrated by an examination of the logic and outcomes of the theoretical, cultural, literary, and other texts and discourses of posthumanism and other managerialisms. As Malm and Hornborg write, anthropocene theory and its "species-thinking on climate change is conducive to mystification and political paralysis."[93] What is necessary to explain and counteract the mystifications of posthumanism and enable social decision in the collective interest is dialectical theory of sociogenic climate change and thus the recognition that what produces these crises is social contradiction—changeable social contradiction which is subject to social decision and planning. Since the start of the Industrial Revolution, it has been the social decision and planning of the class controlling the means of production and distribution which determined not only the forms of consumption and set in place the conditions for the reproduction and exploitation of the labor-power of the working class, but also set in motion the concrete, material changes in the global atmosphere, not only as quantities of pollution but as what climate scientists have now long recognized as a qualitative change in the atmosphere and thus its effects on the natural processes of heat exchange and the like. The natural phenomenon of the greenhouse effect, which is a material precondition for the evolution of human life and thus the development of society, changes through an instance of sociogenic global warming, and the consequences of this change are the effect both of the contradiction of society and nature, but also of the social contradiction within global capitalism itself, such that the ruling class decides what is necessary for its own gain while the working class is largely left out of the processes of decision. Unlike metaphysical speculative theories, which regard contradictions as formal and abstract features of thinking, dialectical explanatory theory grasps contradictions as concrete. In a classic essay on materialist dialectics, Sean Sayers writes that the contrast between metaphysics and dialectics is not merely formal or abstract; rather, "Philosophy and logic can never replace the need for a detailed investigation of the concrete and particular conditions under study."[94] That is, theory is not only a matter of knowledge but "functions as a guide to further thought and action."[95] Metaphysical theory is an ideological guide to social decision, causing the short-term interests of the capitalist class to appear as the commonsense of the species; dialectical materialism is the counter-ideological and concrete theory of historical contradiction which produces the positive knowledge necessary to make social planning and decision into a class-conscious process of sublation and transformation. The resolution of social contradictions and indeed of sociogenic crises such as climate change requires planning. This will either be the managerial planning of the ruling class represented through metaphysics and ideology as the only possibility, or it will be the transformative planning and decision of the collectivity of working people which ends class and thus works to resolve contradictions and mediate crises in the general interest.

Materialist dialectics is necessary to cut through the ideology of posthumanism and thus to provide the explanatory knowledge of concrete reality necessary for ending class, overturning capitalism, and establishing new global relations for the collective good.

BIBLIOGRAPHY

Anderson, Kevin. "Duality in Climate Science." *kevinanderson.info*. October 2015. http://kevinanderson.info/blog/duality-in-climate-science/.

Braidotti, Rosi. *The Posthuman*. Malden, MA: Polity, 2013.

Casarino, Cesare, and Antonio Negri. *In Praise of the Common: A Conversation on Philosophy and Politics*. Minneapolis: University of Minnesota Press, 2008.

Chakrabarty, Dipesh. "The Climate of History: Four Theses." *Critical Inquiry* 35, no. 2 (Winter 2009): 197–222.

Dolphijn, Rick, and Iris van der Tuin, eds. *New Materialism: Interviews and Cartographies*. London: Open Humanities Press, 2012. http://quod.lib.umich.edu/cgi/p/pod/dod-idx/new-materialism-interviews-cartographies.pdf?c=ohp;idno=11515701.0001.001.

Gonzalez, Marcial. "Jameson's 'Arrested Dialectic': From Structuralism to Postmodernism." *Cultural Logic* 2, no. 2 (Spring 1999). http://clogic.eserver.org/2-2/gonzalez.html.

Hardt, Michael, and Antonio Negri. *Empire*. Cambridge, MA: Harvard University Press, 2000.

Jameson, Fredric. *Postmodernism; or, The Cultural Logic of Late Capitalism*. Durham, NC: Duke University Press, 1991.

---. *Representing "Capital": A Reading of Volume I*. New York: Verso, 2011.

---. *Valences of the Dialectic*. New York: Verso, 2009.

Lesjak, Carolyn. "Reading Dialectically." In *Literary Materialisms*, edited by Mathias Nilges and Emilio Sauri, 17–47. New York: Palgrave Macmillan.

Malm, Andreas. "The Anthropocene Myth." *Jacobin* March 30, 2015. https://www.jacobinmag.com/2015/03/anthropocene-capitalism-climate-change/.

Malm, Andreas, and Alf Hornborg. "The Geology of Mankind? A Critique of the Anthropocene Narrative." *The Anthropocene Review* 1, no. 1 (2014): 62–69.

Marx, Karl. *Capital: Volume I*. New York: Penguin Classics, 1990.

---. *The Communist Manifesto*. Edited by Frederic L. Bender. New York: Norton, 1988.

---. *Grundrisse*. New York: Penguin, 1993.

---. "Preface to *A Contribution to the Critique of Political Economy*." In *The Marx-Engels Reader*, 2nd ed., edited by Robert C. Tucker. New York: W. W. Norton & Co., 1978. 3–6.

McEwan, Ian. Interviewed by Friends of the Earth. *Ian McEwan on Climate Change and "Solar" (2010-03-05)*. YouTube video. Last modified March 5, 2010. https://www.youtube.com/watch?v=JXPaKf6u4xI.

---. *Solar*. New York: Doubleday, 2010.

Rees, John. *The Algebra of Revolution: The Dialectic and Classical Marxist Tradition*. New York: Routledge, 1998.

Sayers, Sean. "The Marxist Dialectic." *Radical Philosophy* 14 (Summer 1976): 9–19.

NOTES

1. Rosi Braidotti, *The Posthuman*, (Malden, MA: Polity, 2013), 68.

2. Ibid., 56.
3. Ibid., 68.
4. Rick Dolphijn and Iris van der Tuin, *New Materialism: Interviews and Cartographies* (Open Humanities Press, M Publishing, 2012), 23.
5. Ibid., 23, 123.
6. Karl Marx, cited in John Rees, *The Algebra of Revolution: The Dialectic and Classical Marxist Tradition* (New York: Routledge, 1998), 96.
7. Karl Marx, *Capital: Volume I* (New York: Penguin Classics, 1990), 102.
8. Rees, *Algebra*, 100.
9. Andreas Malm and Alf Hornborg, "The Geology of Mankind? A Critique of the Anthropocene Narrative," *The Anthropocene Review* 1, no. 1 (2014): 64.
10. Ian McEwan, interviewed by Friends of the Earth, *Ian McEwan on Climate Change and "Solar" (2010-03-05)*, YouTube video, last modified March 5, 2010, https://www.youtube.com/watch?v=JXPaKf6u4xI.
11. Ibid.
12. Ibid.
13. Ibid.
14. Ibid.
15. Dipesh Chakrabarty, "The Climate of History: Four Theses." *Critical Inquiry* 35.2 (Winter 2009): 209.
16. Ibid., 210.
17. Ibid., 211.
18. Andreas Malm, "The Anthropocene Myth," *Jacobin*, March 30, 2015. https://www.jacobinmag.com/2015/03/anthropocene-capitalism-climate-change/.
19. Malm and Hornborg, "Geology," 66.
20. McEwan, interviewed by Friends of the Earth.
21. Ian McEwan, *Solar* (New York: Doubleday, 2010), 109.
22. Ibid., 110.
23. Ibid., 109.
24. Ibid., 280.
25. Ibid., 117.
26. Chakrabarty, "Climate," 222.
27. Malm and Hornborg, "Geology," 66.
28. McEwan , *Solar*, 172.
29. Michael Hardt and Antonio Negri, *Empire*, (Cambridge, MA: Harvard University Press, 2000), 237.
30. Cesare Casarino and Antonio Negri, *In Praise of the Common: A Conversation on Philosophy and Politics* (Minneapolis: University of Minnesota Press, 2008), 98–99.
31. Ibid., 199.
32. Ibid., 215.
33. Casarino and Negri, *Praise*, 116–21.
34. . Hardt and Negri, *Empire*, 293.
35. Ibid., 294.
36. Karl Marx, "Preface to *A Contribution to the Critique of Political Economy*," in *The Marx-Engels Reader*, 2nd ed., ed. Robert C. Tucker (New York: W. W. Norton & Co., 1978), 5.
37. Chakrabarty, "Climate," 212, 217.
38. Casarino and Negri, *Praise*, 199.
39. Ibid., 68.
40. Marx, *Capital*, 102.
41. Casarino and Negri, *Praise*, 68.
42. Ibid., 68.
43. Ibid., 68.
44. Ibid., 109.
45. Ibid., 116–21.
46. Ibid., 122.

47. Ibid., 120.
48. Ibid., 86–87.
49. Ibid., 103.
50. Ibid., 121.
51. Ibid., 200.
52. Ibid., 203.
53. Ibid., 203.
54. Ibid., 214.
55. Ibid., 215.
56. Ibid., 216.
57. Ibid., 216.
58. Hardt and Negri, *Empire*, 212–14.
59. Casarino and Negri, *Praise*, 121.
60. Ibid. 122–23.
61. Ibid., 123.
62. Ibid., 123.
63. Karl Marx, *Grundrisse* (New York: Penguin, 1993), 100–101
64. Casarino and Negri, *Praise*, 123.
65. Marx, *Grundrisse*, 100.
66. Ibid., 100.
67. Ibid., 100.
68. Ibid., 101.
69. Fredric Jameson, *Postmodernism; or, The Cultural Logic of Late Capitalism* (Durham, NC: Duke University Press, 1991), 48.
70. Ibid., 51.
71. Fredric Jameson, *Valences of the Dialectic* (New York: Verso, 2009), 410–34.
72. Ibid., 359–60.
73. Fredric Jameson, *Representing "Capital": A Reading of Volume I* (New York: Verso, 2011), 131–32.
74. Ibid., 132.
75. Ibid., 132.
76. Ibid., 133.
77. Ibid., 134.
78. Ibid., 135.
79. Ibid., 137.
80. Ibid., 137.
81. Ibid., 137.
82. Frederic Jameson, *Valences of the Dialectic* (New York: Verso, 2009), 357.
83. Ibid., 357–58.
84. Marx, *The Communist Manifesto*, ed. Frederic L. Bender (New York: W. W. Norton & Co., 1988), 65.
85. Ibid., 65.
86. Jameson, *Valences*, 421.
87. Marcial Gonzalez, "Jameson's 'Arrested Dialectic': From Structuralism to Postmodernism," *Cultural Logic* 2, no. 2 (Spring 1999), http://clogic.eserver.org/2-2/gonzalez.html.
88. Jameson, *Valences*, 421.
89. Carolyn Lesjak, "Reading Dialectically," *Literary Materialisms*, ed. Mathias Nilges and Emilio Sauri (New York: Palgrave Macmillan, 2013), 35.
90. Ibid., 34.
91. Marx, "Preface," 4.
92. Rees, *Algebra*, 64.
93. Malm and Hornborg, "Geology,"67.
94. Sayers, 18.
95. Ibid., 18.

IV

Disaster Theory

EIGHT

The "Event-al" Logic of Disaster and "Left" Extinctionism

Jennifer Cotter, Kimberly DeFazio, Robert Faivre, Amrohini Sahay, Julie Torrant, Stephen Tumino, and Rob Wilkie

Posthumanism operates as a mode of disaster theory that "destructs" humanist thinking while leaving class relations intact. Advocating a cynical politics of "thanaticism,"[1] in which capitalism is rendered as an untransformable ontology from which humanity can no longer escape its inevitable "end," posthumanism focuses on the conceptual negation of the human in the development of new lexical topoi that hover over the tropics of disaster—"extinction events," "despoilment," "mutation events," etc.—but which work to bury causes in effects and construct the social as the singularities of "events" rather than a totality of relations, or materially grounded connections. Disaster theory, in other words, is ruling class theory which, in an apocalyptic tone, deploys disasters—ecological, social, political disasters, which it takes to be allomorphs of nature—to displace history as the outcome of class struggles.

Disaster Theory

Disaster is the tropology by which bourgeois theory invokes a "climate change" in thinking in order to reorient knowledge in the posthumanities. But, "climate" here has nothing to do with the current ecological climate, which is an effect of (the exploitation of) human labor. A theoretical prevarication, the sheer fascination with climatic "ends" in works such as Žižek's *Living in End Times,* Morton's *Hyperobjects: Philoso-*

phy and Ecology at the End of the World, Scranton's *Learning to Die in the Anthropocene*, Thacker's *In the Dust of this Planet*, and Massumi's *The Power at the End of the Economy*, among many others, is merely a reprocessing of the exhausted concepts of recent cultural theory by using marketing techniques that sensationalize the "new"-ness of a present and manufacture the "old"-ness of the past. As Tom Cohen states, "the explication of ecocatastrophic logics, accordingly, are not found in Foucault nor, surprisingly, Derrida,"[2] while as Mackenzie Wark declares, disaster is the idea that "this is not the end of pre-history that the main currents of critical theory thought they would encounter. So perhaps we need some new critical theory."[3] "Climate change" is, in this context, part of a long tradition in bourgeois theory—from Nietzsche's *Beyond Good and Evil* and Freud's *The Ego and the Id* to Derrida's *Of Grammatology* and Butler's *Bodies that Matter*—that questions the naturalness of the natural (in order to represent it according to the needs of capital) while grounding itself as nature, as the uncanny and the unrepresentable with a singularity of its own. For instance, it "deconstructs" politics (which it represents as naturalized beliefs), displacing it into "the political" (as the "endlessness of politics" which includes all life practices, the politics of means without end). "The political" is subsequently also de-naturalized (deprogrammed), but like its textuality, it has the organic unplannedness and unrepresentability of difference—as nature. Nature is the horizon of concept-tropes in "climate change" because it is assumed to be omnihistorical and cross-class, and thus the condition for an environmentalism without class. Thus the refusal "of supposed redemptive 'outsides' to capitalism"[4] and the reassurances that it is not possible to "place the blame at the feet alone of an accidental and evil '1 percent' of corporate culture alone, since an old style revolutionary model does not emerge from this exitless network of systems."[5] The appeal of ecology for bourgeois theory is that a dematerialized, dehistoricized, and depoliticized (and therefore essentially meaningless) "nature" becomes the metonym of "capital" that involves all people, but now *in spite* of class and without end. Instead, "disaster" is about creating a posthuman ethics driven by a posthistorical and cross-class alliance built upon "the narrative about the impending death of the human population, i.e., about the extinction of the human species."[6] It is through the ethics of "ecology" that bourgeois theorists can express concern for the planet and its inhabitants without having to account for the social relations of production which actually shape the ecological. The "end" of nature, in other words, is really the continuation of capital.

As part of its new lexical topoi, disaster theory displaces the critique and transformation of historically produced social relations and structures with what it calls "deep history" which is an ideological repackaging of sociobiology and so-called "evolutionary psychology" and the idea, as John Gray advocates, that "the destruction of the natural world is not the result of global capitalism, industrialisation, 'Western Civilization,' or any flaw in human institutions. It is the consequence of the evolutionary success of an exceptionally rapacious primate."[7] Nowhere is this more evident than in the "deep ecology" or "deep history" trope of the "anthropocene." Anthropocene is the trope of climate change without class relations. It alibis capitalist social relations by attributing crises—from economic to ecological crises—that are brought about by human intervention *under historically produced social relations of production* that prioritize the profit of some over the needs of all, to *species-wide* traits developed in the early evolution of humans. A symptomatic example is Claire Colebrook's book *Death of the PostHuman* in which she claims that climate change is owing to humanity's "internal and ever-present . . . horrific tendency towards self-extinction."[8] Colebrook attributes a range of effects of the social division of labor and private property relations to species-wide evolutionary traits of *homo sapiens.* Colebrook claims

> Just as the very desire for fats and sugars that propelled the body to hunt and develop technologies for metabolic stability and survival will drive the modern body into obesity, hypertension and an early grave, so the darting eye that stimulated the brain into becoming a reading and interpreting animal, may also be at the forefront of the human species' cognitive atrophy.[9]

Thus, in the "anthropocene" narrative social and material contradictions such as the rise of widespread obesity and hypertension among the U.S. and U.K. workforces, climate change and the pollution of the environment, the deterioration and suppression of public critique, and the cynical reduction of workers by capital to "targeted consumers of dazzling spectacle"[10] are all ideologically cleansed of their relationship to the material relations of capitalism and attributed instead to "self-extinguishing tendencies"[11] of the human species and "the brain's own capacity to destroy itself through the very perceptive power that generated its supposedly proper potentiality in the first place."[12] What is posited here in other words is human "destiny"—which in a reversal of humanism, rewrites human progress as human devolution and species suicide.

According to Colebrook, it is the "drive for life"—that is, the evolutionary survival of the species—that is at the root of the destruction of biodiversity.[13] Thus, Colebrook cynically implies, *human extinction* is the only viable solution to anthropogenic climate change: "What might be thought is the extinction of the climactic eye: can we imagine a mode of reading the world, and its anthropogenic scars, that frees itself from folding the earth's surface around human survival?"[14]

Colebrook's argument is a version of what Andreas Malm and Alf Hornborg, in their essay "A Geology of Mankind? A Critique of the Anthropocene Narrative," call the "anthropocene displacement":

> climate change is *denaturalised* in one moment—relocated from the sphere of natural causes to that of human activities—only to be *renaturalised* in the next, when derived from an innate human trait, such as the ability to control fire. Not nature, but human nature—this is the Anthropocene displacement.[15]

Yet "Realising that climate change is '*anthropogenic*,'" Malm and Hornborg argue, "is really to appreciate that it is *sociogenic*," [16] by which they mean that climate change occurs within specific social relations of production and is historical, not natural, or an effect of a species-wide trait. Steam engines, for example, "were not adopted by some natural-born deputies of the human species" but rather, on the contrary, "by the nature of the social order of things, they could only be installed by the owners of the means of production."[17] In the logic of "anthropocene," "deep ecology," and "disaster" narratives, the fact that "the historical origins of anthropogenic climate change were predicated on highly inequitable global processes from the start" is dissolved into the ebb and flow of biodiversity and the inevitable evolution and extinction of species.[18] Climate change, however, is not a "species" issue (that is, ahistorical and post-class), but a class issue: it is historical and an effect of what happens at the point of production.

In posthumanist disaster theory, "climate change" is translated into the vulgar, apocalyptic, and loudly sentimental version of such other "terrestrial" theories as "risk society theory."[19] Since hazards, like all environmental events, are represented as affecting all people alike—the poor and the rich—the division of the social into owners and workers is no longer relevant to the risk society. Of course, this does not mean that risk theory does not recognize that poor neighborhoods are more likely to suffer from pollution, contaminated water, and proximity to the toxic waste of the affluent. It knows the politics of hazard, but treats it as a local matter that does not substantively influence the terrestrial reach of the theory. Having erased class from the social, risk theory invokes a messianic voice (which is also the tone of "climate change") to claim that old social models of analysis concerned with equality are outdated, modernist political thought and residues of industrial class society. In the new risk society, "safety" not "class" is the norm of social justice.

Disaster theory represents "hazard" and "risk" as metaphysical conditions of being abstracted from relations of exploitation. At a time when earlier celebratory ideologies of "New Times," "post-Fordism" and a "fundamental break" in capitalism to a so-called "post-capitalist society" have been thrown into sharp relief as blatant forms of neoliberal ideology by the ongoing economic crisis and intensifying class contradictions in capitalism, disaster theory posits what Timothy Morton in his book *Hyperobjects* calls a "fundamental shaking of being" or "*being-quake*"[20] that (again, ever again!) renders the critique of capitalism "outdated."[21]

Disaster theory, in other words, renews the old ruling class ideology of a fundamental break in capitalism by installing it at a new level of metaphysical abstraction: as an ontological shift within the structure of reality as such.

Deploying the emergent discourses of object oriented ontology, speculative realism, and so-called "new materialism," the posthumanities now posit "disaster" as a fundamental ontological condition of being. Disaster theory proposes that there exists a "wholly other"[22] "world-without-us"[23] made up of an "other" dark reality: a world of irreducible "hyperobjects,"[24] in which "each thing is an inexhaustible surplus"[25] that exceeds our cognitive abilities and lie outside of the historical series of social struggles over material relations. According to the ontology of disaster, our physical encounters and "crashes" into this unfathomable, autonomous, and alternate reality have shattered the world as we know it:

> There we were, trolling along in the age of industry, capitalism, and technology, and all of a sudden we received information from aliens, information that even the most hardheaded could not ignore, because the form in which the information was delivered was precisely the instrumental and mathematical formulas of modernity itself.[26]

Disaster theory ideologically dissolves and trivializes, in the language of "alien invasion," the social and economic crisis of capitalism and the devastating consequences that the exploitation of humans by humans continues to wreak on both the human and nonhuman world. It displaces the social struggle for critique and transformation of the world-historical social relations of production in capitalism with proclamations of the "end of the world" and the idea that at the world's end, critique and collective transformation are now outdated relics of the Anthropocene era. "End of the world" narratives, like all "end" narratives—the end of history, the end of ideology, the end of socialism, etc.—are grounded in a class theology the purpose of which is to represent existing social relations and practices as outside the reach of history and, therefore, as immune from critique and collective, material, social transformation (i.e., "praxis," that taboo term of the posthumanities).

The theoretical hollowness of "critical climate change" is nowhere more on display than in the treatment of disaster as an "event" (e.g., Morton's "end of the world"). "Event" is the trope of the post-historical. It "implies surprise, exposure, the unanticipatable"[27] and its "eventfulness depends" on the "experience of the impossible" which is the "unique, exceptional, and unpredictable arrival of the other."[28] But, and here is the event's political twist, as "there is iterability and return in absolute uniqueness and utter singularity, [this] means that the arrival of the arrivant—or the coming of the inaugural event—can only be greeted as a return, a coming back, a spectral *revenance*."[29] The "event," in cultural theory is the causeless, reversible history, utter *alea*; its arrival is always a re-arrival outside the regulative structure of specific historical social relations. The deploying of "event" as the analytics of "climate change" is a symptom of the poverty of its historical understanding masquerading as a freedom from connected meanings (history).

"Event-al" thinking is valorized because it is the paralogic of singularities that rewrites history as an uninterrupted series of irruptions without causality. Cultural theory's "climate change" is the administrative form of this historical disconnection. By severing the cause (capitalism) from its effects (climate change) and by attributing such effects to the "absent cause" of modernity's legacies, "event-al" thinking allows people to believe that there is nothing to be done, no social(ist) project, to halt the environmental crisis. Eco-ontology, a conception of being-in-common, is thus put forward as an ethical solution to what are taken to be socially irresolvable contradictions (climate change, global inequality . . .). Take inequality for instance: Inequality is an effect of exploitation, it is the appearance of the social relations through which capital exploits labor by the extraction of surplus-value in the workday. In the "object-oriented-ontology," however, class as exploitation is de-centered as "hyperobject" and mystified as "a whole spatio-temporal geography [. . .] a network structure, mesh, or ecology" in which "the places where we live, the manner in which roads, public transportation, and infrastructure are organized, the availability of jobs, linguistic dialectics into which one is born, etc., take on a life of their own, structuring and organizing human relations."[30] If "class" exists at all, in this view, it is the contingency of "what I have in common with others" or the multiple overlaps of "different classes, reduced or enlarged—classes of inheritance, of ideas, of thought, of belief, and of action" that include both the human and the nonhuman.[31] It is this being-in-common of the human and the nonhuman which supposedly "ruins the universe of accumulated objects and events, with either more or less intensified values" and brings with it the recognition of formal equality in which any individual "being" is "no more in the world than the smallest dust particle, than a fraction of this dust particle, than the word 'dust,' than language, than a galaxy, than a supergalaxy, than a contradiction, or than any part of [itself]."[32] Through the singular, "event-al" thinking produces subjectivities out-of-joint—subjectivities who are alienated from grasping the historical, and therefore collectively transformable, basis of their conditions of life—and who are, therefore, only prepared for desire-al shopping in the alea of the global marketplace but now with a renewed ethical sense of purpose in their consumption. Class, in other words, is on one hand ideologically translated into a multi-valenced "thing"—a hyper-object, an ecology, a way of life—that is "non-sublatable" rather than an historical relation of exploitation that needs to be abolished, while, on the other, it is rendered ontologically "equivalent" ("no more in the world") to every other "object" and thus without any determining priority. The "ecological" solution to class inequality is therefore to equalize access to markets in which the "objects" that make up "class" (speech patterns, where one lives, quality of roads . . .) are more freely available, while the basic class arrangements through which capital extracts surplus-value from labor is occulted and thereby silently maintained. Disaster theory acknowledges class inequality—as a global "disaster"—while proposing survival strategies within the disaster (regional changes) over fundamental social change—strategies which above all render abolishing class relations impossible. "Climate change" is the pedagogy of a post-disaster longing. It teaches workers the trauma of the rationalities of modernity and re-educates them into spontaneity so that they can shop (beyond reason) to slow capital's falling rate of profit (as absolute reason).

Disaster theory is the cultural arm of what Naomi Klein in her *The Shock Doctrine* called "disaster capitalism." It is an ideological crusade to invert cultural meanings, which are the outcome of specific historical social relations of production, into the spectral wavering of mutational signs (*l'animot*, ecomimesis . . .), in order to make it impossible to form any ground for class struggles. Klein argues that capitalism uses disasters to heighten cultural crises and produce a "climate change" within which it pushes through laissez-faire, pro-corporation policies as "new" ideas. However, Klein's theory of "disaster capitalism" is not a materialist critique of capitalism but a "new materialist" theory for a "green capitalism." In her theory of "disasters," the "disasters" of climate change and global inequality are socially engineered by neoliberals to expand markets through the privatization of the common(s). In other words, certain ideas and interests ("neoliberalism" or "free market fundamentalism"[33]) have acquired the power to shape the social over those of others. Behind the world we see, she suggests, is the spectral world of ideology that makes it. The question her new materialist theory of capitalism cannot answer is, what has empowered these ideas? Why neoliberalism now? To answer this question entails grasping capitalism as a totality, an integrated understanding of how the "base"-ic class relations of exploitation and the economic laws that follow from them shape the cultural superstructure. The drive for markets—neoliberalism—is not simply an (un)ethical policy that can be corrected by a (re)turn to a "better" capitalism. Rather, it is an objective outcome of the structural contradictions of capitalism and its resulting falling rate of profit that cannot be ended without ending class relations. The drive for markets and privatization are themselves an outcome of the falling rate of profit as explained by Marx: the fact that in order to capture more surplus-value "capital must nestle everywhere, settle everywhere, establish connexions everywhere"[34] in order to break out of its periodic crisis of overproduction.

What has "changed everything" is not the terraforming of the earth by narrow-minded economic ideas, as Klein (*This Changes Everything: Capi-*

talism vs. the Climate), Morton (*Hyperobjects: Philosophy and Ecology after the End of the World*), and now even the Pope argue. Indeed, Pope Francis' framing of "disaster" mirrors exactly the linguistics of "climate change" in the posthumanities. He argues that climate change follows from "modern anthropomorphism" because of how it "sees nature as an insensate order, as a cold body of facts, as a mere 'given,' as an object of utility."[35] The response to climate change therefore requires a new "cultural ecology," a movement away from the "throwaway culture" of modernity to an "ecological culture" in which we "pay renewed attention to reality and the limits it imposes."[36] To ignore the "limits" reality imposes on human abilities "would be nothing more than romantic individualism dressed up in ecological garb, locking us into"[37] a "Promethean vision of mastery over the world."[38] In short, a "belief" in human-centeredness is said to be outdated at a time of global environmental crisis, but rather than change the basic class arrangements that explain the (mis)use of nature for profit over meeting human needs in an environmentally sustainable mode, a "new" ecological belief system is proposed that makes change into a change of "heart." According to Klein, Pope Francis' encyclical on climate change is "right on the economics."[39] This is because the Pope's "economic" theory is the same as Klein's "new materialist theory" of capitalism. Class is resignified from being the basic antagonism inscribed in the work day between exploiters and exploited into a culture war between different "beliefs" (humanists vs. posthumanists). To be sure, the Pope opposes "biocentrism" and yet believes in the inherent "dignity" of humans that authorizes their "dominion" over nature,[40] but this cultural difference should not distract from the underlying class ideology and interests that posthumanism advances beyond the regional idiomatic expressions it takes. Posthumanism always inverts material causes into imaginary (spiritual) agencies and thereby alibis capitalism.

Change in the posthumanist turn amounts to an event-al change in "belief" (from humanist to posthumanist) and culture is trivialized as the "sensorium" where the "encounter" between the human and its other is "affected" ("throwaway culture" vs. "ecological culture"), rather than the ideological zone where the classes "become conscious of their conflict and fight it out."[41]

Class is the Root of Crisis

"Society," Marx writes, "is the complete unity of [humanity] with nature"[42] because "nature is [humanity's] *inorganic body*."[43] The social is the expression of the way in which humanity produces itself and, in turn, produces knowledge of itself and the natural world around it. The "social" in other words is the organization of individuals in the collectivity of labor and, in turn, "the individual *is the social being*."[44] Posthumanism's turn toward *being-in-common*—which displaces the contradictions of society in the embracing of a posthistorical, postbinary, postdialectical idealism of a "planetary" consciousness—is the erasure of class relations and the rendering of inequality as conceptually dispersed and ungraspable. Whereas posthumanists are willing to be historical, but only in terms of the genealogy of ideas about nature (e.g., Derrida's *The Animal That Therefore I Am*, Agamben's *The Open: Man and Animal*, Simondon's *Two Lessons on Animal and Man . . .*), to approach the contradictions of the contemporary historically, such that one can not only understand but seek to transform them, means beginning with the material realities of life as they are shaped and reshaped by the mode of production. As Marx explains,

> The *distinction* between capital and land, between profit and rent, between both and wages, and *industry*, and *agriculture*, and *immovable* and *moveable* private property—this distinction is not rooted in the nature of things, but is a *historical* distinction, a *fixed* historical moment in the formation and development of the contradiction between capital and labor.[45]

The commercialization of the environment and the environmental crisis is, in other words, not an ontological or epistemological failure, but an effect of the fundamental logic of capitalism which began with the commodification of labor at its base that changed the relation of humans to nature from "subsistence" (production for use) to "exploitation" (production for exchange). Prior to capitalism there was no economic incentive to systematically develop the forces of production and increased surpluses of labor were tied to the limits of subsistence agricultural production under feudal social relations. It was only when the bourgeois revolution dispossessed the peasants of the land thereby making them into "free" wage-laborers with nothing to sell on the market but their ability to work that there arose a social incentive to augment technology to increase the productivity of labor and the law of profit and capital accumulation became the rule. The commodification of nature, to say it again, is not due to a cultural regime of "ideas" (anthropocentric humanism, market fundamentalism . . .), but the logic of capitalism inscribed in wage-labor relations. If capital must today confront the effects of the techniques of production used today to extract surplus-value from labor, it does not mean, as argued by Negri, that capitalism has ceased to be based upon the "law of value" and instead has become a "law of command" over the multitude.[46] Rather, capitalism will seek out and employ ideas which expand the conditions of exploitation in ways that may bring different segments of capital into conflict with one another—fossil fuel companies versus transportation companies, technology companies versus energy companies, etc.—but none will propose eliminating the fundamental contradiction between capital and labor. Hence, the rise of "disaster theory" and advocacy, at best, of "green capitalism." These are theories for managing contradictions, for opening up "new" ways of thinking about how to best organize production for profit, but in no way do they challenge the exploitative relations fundamental to capitalism. In fact, they propose that to do so is too "human-centered" and therefore impossible. The exploitative relation of humans to nature, however, will only change from a means for increasing profits at the expense of meeting and augmenting human needs through a global proletarian revolution that assumes power and brings about a "society [that] inscribes on its banner: From each according to his ability, to each according to his needs!" as the rule.[47] This entails theorizing the relation of humans to nature as a material question that is at root grounded in class relations rather than merely a cultural matter about people's beliefs. Making the politics of climate change into a cultural war over beliefs only empowers the already powerful who control the production and dissemination of ideas whose class interest requires the reforming of capitalism the better to maintain it, and their positions in it, at a time when it is obviously failing. The sign of the bourgeois hijacking of the environmental movement is nowhere more clear than in how "climate change" has been turned into further ludic text-wares.

Disaster theory produces an epistemological crisis—"climate change"—in which meaning as social relations is represented as limited and unable to engage the higher levels of "extinction events." This is another way of saying that through the invocations of disaster, cultural signs are re-signified so as to produce meanings that, instead of unconcealing social relations and laying bare their class logic, become linguistic disasters to the body politics and put its class interpretations of signs in ruins. Red cultural theory and critique produce public meanings through which social collectivities are formed and general class experiences (class-in-itself) are transformed into class consciousness and class solidarity (class-for-itself). Cultural theory and critique, in short, have "use-value." Disaster theory undoes their use-values and reconstitutes theory and critique as commodified exchange-values for profit. The public meaning of texts (the collective signified) is no longer a means for analyzing and understanding the social in order to change it. Reading, consequently, is no longer aimed at understanding the outside of the text but is a purely immanent activity directed toward the anarchograms that put the signified in ruins. Reading is the disaster of meaning, a semiotic catastrophe, a mutation of the letteral. The bigger the disaster, the more radical is the re-signification. The ideal disaster is a mega-disaster with a "postglobal" magnitude moving in the rhythm of the "end of the world." Disaster theory is the trauma of the public signified. It softens it up for lines of private connotations that ultimately subject class to affective affinities. In this way, disaster theory is the control of the archive: the exclusion of class and reassertion, with "climate change," of the rights to private ownership of public surplus labor. But all archives are historical—they are the scene of class struggles. Their control will collapse not because they are archiviolithic—the lapse of memory and forgetting—but because the class contradictions that they attempt to contain inevitably break through the surface.

To understand the diverse dissemination of "crises" today as well as their ever "newer" theoretical "solutions" one must seek out the connecting roots underneath the surfaces of culture and politics, namely the class relations which shape the totality of contemporary society. Doing so requires an approach that is the direct opposite of bourgeois theory, the assemblage of reading practices that bury causes in effects and construct the social as singularities of events, as causeless arrivals. In a Nietzschean move repeated time and again in "new" idioms, bourgeois theory suspends the very idea of "cause" in the inscription of bodies by the "wholly other." Bourgeois theory, to use Marx and Engels' words, has always been "instruction in the art of ghost-seeing,"[48] or the metaphysics of capitalism. It has conjured up concepts, tropes, interpretive strategies and a rhetoric to go with them, by which capital, often in the stylistic mode of an anarcho-left, replaces property with the power (of the State), inverts class difference into cultural difference, obscures the collective in irruptive singularities, and legitimates its own class interest as the universal interest. Faced with an intensification of class contradictions—whose stubborn materialist complexities overflow the interpretive cunning of its textual ethics and threaten to make it irrelevant to capital and deprive it of its institutional rewards—bourgeois theory is now altering and adapting its strategies in order to translate the class interests of capital into the new topoi of posthumanism through which it distracts analytical attention away from material causes and instead focuses on cultural effects. In contrast, moving beyond the tropic politics of "disaster" requires, as Marx says, "distinguish[ing] the language and imaginary aspirations of parties from their real organism and their real interests."[49] While left bourgeois theory acts as a medium for conjuring up ghostly explanations for the disasters of capitalism that deflect attention from class exploitation, what is necessary today if we are to create a world that meets the needs of all is the scientific praxis of red critique.

BIBLIOGRAPHY

Agamben, Giorgio. *The Open: Man and Animal.* Translated by Kevin Attell. Stanford, CA: Stanford University Press, 2004.

Beck, Ulrich. *Risk Society: Towards a New Modernity.* London: Sage Publications, 1992.

Bryant, Levi R. "On the Reality and Construction of Hyperobjects with Reference to Class." *Speculations* II (2011): 86–103.

Butler, Judith. *Bodies That Matter: On the Discursive Limits of "Sex."* New York: Routledge, 1993.

Cohen, Tom. "Introduction: Murmurations—'Climate Change' and the Defacement of Theory." In *Telemorphosis: Theory in the Era of Climate Change,* vol. 1, edited by Tom Cohen, 13–42. Ann Arbor, MI: Open Humanities Press, 2012.

Colebrook, Claire. *Death of the Posthuman: Essays on Extinction*, vol. 1. Ann Arbor, MI: Open Humanities Press, 2014.

Derrida, Jacques. *The Animal That Therefore I Am*. Translated by David Wills. Edited by Marie-Louise Mallet. New York: Fordham University Press, 2008.

---. "A Certain Impossible Possibility of Saying the Event," *Critical Inquiry* 33, no. 2 (2007): 441–61.

---. *The Gift of Death*. Translated by David Wills. Chicago, IL: University of Chicago Press, 1995.

---. *Of Grammatology*. Translated by Gayatri Chakravorty Spivak. Baltimore, MD: Johns Hopkins University Press, 1976.

Francis, *Laudato Si': On Care for our Common Home*. Vatican City: Libreria Editrice Vaticana, 2015.

Freud, Sigmund. *The Ego and the Id*. Translated by Joan Riviere. New York: W. W. Norton & Co., 1962.

Garcia, Tristan. *Form and Object: A Treatise on* Things. Translated by Mark Allan Ohm and Jon Cogburn. Edinburgh, Scotland: Edinburgh University Press, 2014.

Gray, John. *Straw Dogs: Thoughts on Humans and Other Animals*. London: Granta Publications, 2002.

Harman, Graham. *Bells and Whistles: More Speculative Realism*. Winchester, UK: Zero Books, 2013.

Klein, Naomi. *The Shock Doctrine: The Rise of Disaster Capitalism*. New York: Metropolitan Books/Henry Holt, 2007.

---. *This Changes Everything: Capitalism vs. The Climate*. New York: Simon & Schuster, 2014.

Malm, Andreas and Alf Hornborg, "The Geology of Mankind? A Critique of the Anthropocene Narrative," *The Anthropocene Review* 1, no. 1 (2014): 62–69.

Marx, Karl. *Economic and Philosophic Manuscripts of 1844*. In *Marx-Engels Collected Works*, vol. 3, 229–346. New York: International Publishers, 1975.

---. *The Eighteenth Brumaire of Louis Bonaparte*. In *Marx-Engels Collected Works*, vol. 11, 99–197. New York: International Publishers, 1979.

---. *Contribution to the Critique of Political Economy*. In *Marx-Engels Collected Works*, vol. 29, 257–417. New York: International Publishers, 1987.

---. *Critique of the Gotha Program*. In *Marx-Engels Collected Works*, vol. 24, 75–99. New York: International Publishers, 1988.

Marx, Karl and Frederick Engels. *The German Ideology*. In *Marx-Engels Collected Works*, vol. 5, 19–584. New York: International Publishers, 1976.

---. *Manifesto of the Communist Party*. In *Marx-Engels Collected Works*, vol. 6, 477–519. New York: International Publishers, 1976.

Massumi, Brian. *The Power at the End of the Economy*. Durham, NC: Duke University Press, 2015.

Morton, Timothy. *Hyperobjects: Philosophy and Ecology after the End of the World*. Minneapolis: University of Minnesota Press, 2013.

Negri, Antonio. *Marx Beyond Marx: Lessons on the Grundrisse*. Translated by Harry Cleaver, Michael Ryan, and Maurizio Viano. Edited by Jim Fleming. New York: Autonomia, 1992.

Nietzsche, Friedrich. *Beyond Good and Evil: Prelude to a Philosophy of the Future*. Translated by R. J. Hollingdale. Harmondsworth: Penguin, 1990.

Scammell, Rosie. "Pope Francis recruits Naomi Klein in climate change battle," *The Guardian (UK)*, June 27, 2015. http://www.theguardian.com/world/2015/jun/28/pope-climate-change-naomi-klein.

Scranton, Roy. *Learning to Die in the Anthropocene: Reflections on the End of a Civilization*. San Francisco: City Lights, 2015.

Simondon, Gilbert. *Two Lessons on Animal and Man*. Translated by Drew S. Burk. Minneapolis, MN: Univocal, 2011.

Thacker, Eugene. *In the Dust of this Planet*. Winchester, UK: Zero Books, 2011.

Wark, Mackenzie. "The Birth of Thanaticsm." *Public Seminar*. April 3, 2014. http://www.publicseminar.org/2014/04/birth-of-thanaticism/.
---. *Molecular Red: Theory for the Anthropocene*. London: Verso, 2015.
Žižek, Slavoj. *Living in the End Times*. London: Verso, 2010.
Zylinska, Joanna. *Minimal Ethics for the Anthropocene*. Ann Arbor, MI: Open Humanities Press, 2014.

NOTES

1. Mackenzie Wark, "The Birth of Thanaticsm," *Public Seminar*, April 3, 2014, http://www.publicseminar.org/2014/04/birth-of-thanaticism/.
2. Tom Cohen, "Introduction: Murmurations—'Climate Change' and the Defacement of Theory," in *Telemorphosis: Theory in the Era of Climate Change Volume 1*, ed. Tom Cohen (Ann Arbor, MI: Open Humanities Press, 2012), 16.
3. Mackenzie Wark, *Molecular Red: Theory for the Anthropocene* (London: Verso, 2015). xii.
4. Tom Cohen, "Introduction: Murmurations", 17.
5. Ibid., 15.
6. Joanna Zylinska, *Minimal Ethics for the Anthropocene* (Ann Arbor, MI: Open Humanities Press, 2014), 9.
7. John Gray, *Straw Dogs: Thoughts on Humans and Other Animals* (London: Granta Publications, 2002), 7.
8. Claire Colebrook, *Death of the Posthuman: Essays on Extinction*, vol. 1 (Ann Arbor, MI: Open Humanities Press, 2014), 19.
9. Ibid., 21.
10. Ibid., 13.
11. Ibid., 11.
12. Ibid., 21.
13. Ibid., 22.
14. Ibid., 23.
15. Andreas Malm and Alf Hornborg, "The Geology of Mankind? A Critique of the Anthropocene Narrative," *The Anthropocene Review* 1, no. 1 (2014): 65.
16. Ibid., 66.
17. Ibid., 63–64.
18. Ibid., 63.
19. See, for instance, Ulrich Beck's *Risk Society: Towards a New Modernity*.
20. Timothy Morton, *Hyperobjects: Philosophy and Ecology After the End of the World* (Minneapolis: University of Minnesota Press, 2013), 19.
21. Ibid., 1.
22. Jacques Derrida, *The Gift of Death*, trans. David Wills (Chicago, IL: University of Chicago Press, 1995), 57.
23. Eugene Thacker, *In the Dust of this Planet* (Winchester, UK: Zero Books, 2011), 5.
24. Morton, *Hyperobjects*, 1.
25. Graham Harman, *Bells and Whistles: More Speculative Realism* (Winchester, UK: Zero Books, 2013), 61.
26. Morton, *Hyperobjects*, 19.
27. Jacques Derrida, "A Certain Impossible Possibility of Saying the Event," *Critical Inquiry* 33, no. 2 (2007), 441.
28. Ibid., 452.
29. Ibid., 452.
30. Levi R. Bryant, "On the Reality and Construction of Hyperobjects with Reference to Class," *Speculations* II (2011), 88.
31. Tristan Garcia, *Form and Object: A Treatise on Things*, trans. Mark Allan Ohm and Jon Cogburn (Edinburgh, Scotland: Edinburgh University Press, 2014), 363.
32. Ibid., 435.

33. Naomi Klein. *This Changes Everything: Capitalism vs. The Climate* (New York: Simon & Schuster, 2014), 64–95.
34. Karl Marx and Frederick Engels, *Manifesto of the Communist Party*, *Marx-Engels Collected Works*, vol. 6 (New York: International Publishers, 1976), 487.
35. Francis, *Laudato Si': On Care for our Common Home* (Vatican City: Libreria Editrice Vaticana, 2015), 86.
36. Ibid., 87.
37. Ibid., 89.
38. Ibid., 87.
39. Rosie Scammell, "Pope Francis recruits Naomi Klein in climate change battle," *The Guardian (UK)*, June 27, 2015. http://www.theguardian.com/world/2015/jun/28/pope-climate-change-naomi-klein.
40. Francis, *Laudato Si'*, 88–89.
41. Karl Marx, *Contribution to the Critique of Political Economy*, in *Marx-Engels Collected Works*, vol. 29 (New York: International Publishers, 1987), 463.
42. Karl Marx, *Economic and Philosophic Manuscripts of 1844*, in *Marx-Engels Collected Works*, vol. 3 (New York: International Publishers, 1975), 298.
43. Ibid., 276.
44. Ibid., 299.
45. Ibid., 285.
46. Antonio Negri, *Marx Beyond Marx: Lessons on the Grundrisse*, trans. Harry Cleaver, Michael Ryan, and Maurizio Viano, ed. Jim Fleming (New York: Autonomia,1992): 171–90.
47. Karl Marx, *Critique of the Gotha Program*, in *Marx-Engels Collected Works*, vol. 24 (New York: International Publishers, 1988), 87.
48. Karl Marx and Frederick Engels, *The German Ideology*, in *Marx-Engels Collected Works*, vol. 5 (New York: International Publishers, 1975), 152.
49. Karl Marx, *The Eighteenth Brumaire of Louis Bonaparte*, in *Marx-Engels Collected Works*, vol. 11 (New York: International Publishers, 1979), 128.

Index

About the Contributors

Jennifer Cotter, PhD, has published essays on feminism, materialism, transnationalism, and the critical (post)humanities in such books as *Stories of Post-Human Cultures* as well as *the minnesota review, College Literature, Textual Practice, Nature, Society, and Thought*, and (in Polish translation) *Lewą Nogą*. She is currently completing her forthcoming book, *Whatever Happened to Materialist Feminism?* She teaches critical theory and world literature at William Jewell College.

Kimberly DeFazio, PhD, is the author of *City of the Senses* (Palgrave 2011) and the forthcoming *Materialism and Resistance*. Her writings have also appeared in such books as *Marxism and Urban Culture* and *Confronting Universalities: Aesthetics and Politics Under the Sign of Globalisation*. She teaches at the University of Wisconsin-La Crosse.

Robert Faivre is Professor of English at Adirondack Community College, State University of New York. His essay "Class: The Base of All Reading" appears in *Class in Education: Knowledge, Pedagogy, Subjectivity* (Routledge 2010). He is writing a book on Fredric Jameson and dialectics of reading.

Amrohini Sahay, PhD, has published scholarly essays and given papers on race, postcoloniality, transnationality, cyberculture, and contemporary theory in such journals as *New Literary History, Cultural Logic, Textual Practice*, and *College Literature*, and at national and international conferences. She is currently working on several projects on "empire," globality, and capitalism and teaches at Hofstra University.

Julie P. Torrant, PhD, is the author of *The Material Family* (Sense Publishers, 2011) and has published articles on feminism, family, and critical theory in such journals as *Signs, Cultural Logic*, and *Textual Practice*. She is currently working on a book about health and disability in global capitalism.

Stephen Tumino, PhD, is the author of *Cultural Theory After the Contemporary* (Palgrave Macmillan 2011). His scholarly writings have appeared in such journals as *Rethinking Marxism, Textual Practice*, and *Nature, Society, and Thought*. Some of his writings have been translated into Persian and Turkish. He currently teaches at the Borough of Manhatten Community College (CUNY).

Rob Wilkie, PhD, is the author of *The Digital Condition: Class and Culture in the Information Network*, published by Fordham University Press (2011) and his essays have appeared in such journals as *International Critical Thought*; *Textual Practice*; *Nature, Society and Thought*; and *Postmodern Culture*. He teaches at the University of Wisconsin-La Crosse.

www.ingramcontent.com/pod-product-compliance
Ingram Content Group UK Ltd.
Pitfield, Milton Keynes, MK11 3LW, UK
UKHW012037150926
475560UK00001B/59

9 781498 505734